To Know the World

To Know the World
A New Vision for Environmental Learning

Mitchell Thomashow

The MIT Press
Cambridge, Massachusetts
London, England

© 2020 Massachusetts Institute of Technology

All rights reserved. No part of this book may be reproduced in any form by any electronic or mechanical means (including photocopying, recording, or information storage and retrieval) without permission in writing from the publisher.

This book was set in Sabon LT Std by New Best-set Typesetters Ltd. Printed and bound in the United States of America.

Library of Congress Cataloging-in-Publication Data

Names: Thomashow, Mitchell, author.
Title: To know the world : a new vision for environmental learning / Mitchell Thomashow.
Description: Cambridge, Massachusetts : The MIT Press, 2020. | Includes bibliographical references and index.
Identifiers: LCCN 2020002975 | ISBN 9780262539821 (paperback)
Subjects: LCSH: Environmental education. | Human ecology. | Nature—Effect of human beings on.
Classification: LCC GE70 .T53 2020 | DDC 363.70071—dc23
LC record available at https://lccn.loc.gov/2020002975

10 9 8 7 6 5 4 3 2 1

What is the single best thing a person can do for tomorrow's world?
—Richard Powers, *The Overstory*

Contents

Acknowledgments   xi

Part I: Why Environmental Learning Matters

1   **The Past and Future of Environmental Learning**   3
     Now More Than Ever   3
     What's Next and Why   5
     The Whole Earth Beckons   5
     A Book Becomes a Field   6
     The Emerging Tides of Change   7
     Environmental Learning and the Tides of Change   9
     A New Generation   10
     The Ecological Knowledge Gap   11
     Thinking Things Through   12

2   **Memory Forever Unfolding**   19
     The Branches of Memory   19
     The Golden Guide to the Stars   20
     Harold and the Purple Crayon   21
     Here Comes the Fog Man   23
     Developmental Interludes   24
     Perennial Learning   26
     Adaptive Learning   27
     An Ocean Voyage   29

Part II: Environmental Learning in the Anthropocene

3   **The Tides of Change**   35
     Pacific Northwest Change Makers   35
     The Pacific Northwest: A Brief Profile   37

Now the Anthropocene 39
An Intricate Convergence 41
Always the Biosphere: Environment 43
Tribes and Territories: Diversity 46
The Stress of Inequality: Equity 48
Participation and Democracy: Inclusion 50
Openness and Vigilance 52
It's the Human Condition, Isn't It? 54

4 **Is the Anthropocene Blowing Your Mind?** 57
Screens, Windows, and Frames 57
Autonomous and Autonic 61
A Collective Spell 63
Ubiquitous Novelty 66
Reclaiming Attention 67
Is Autonomy Possible? 69
Clarifying Intention 71
How Can Thoreau Help? 73
Reading the Day 75
Passing Moments of Environmental Learning 76
Proliferation and Filters 78
Collect and Consume 79
Curate and Connect 81

Part III: A New Vision for Environmental Learning

5 **Constructive Connectivity** 87
Visiting the Golden Spike 87
Network Observation Interlude 90
Biosphere Network Archetypes 93
Ecological Networks and Their Visualization 98
Visualizing the Biosphere 101
Social Networks 103
Networks, Identity, and Education 105
Learning to Navigate Networks 107
Constructive Connectivity Continued 111

6 **Migration: The Movement of People and Species** 113
Family Stories 113
Geno 2.0: Do You Know Where Your Ancestors Are? 115
Refugees, Migrants, and Immigrants 117

Why Environmental Learning about Migration Matters 120
Migration and the Biosphere 122
The Original Human Globalization 124
Converging Migrations and the World City 127
The Genomic Legacy 129
Empathy, Compassion, and Education 131
Mapping Migration: From Biosphere to Psyche 132

7 **Cosmopolitan Bioregionalism** 137
Maps, Patterns, and Places 137
The Original Bioregionalists 140
Borders, Boundaries, and Ecosystems 142
Tribes, Territories, and Sovereignty 145
Why Place Still Matters 148
Why Bioregionalism Still Matters 150
The Biosphere and the City 152
Contact Zones: Living with Difference 155
So What Is Cosmopolitan Bioregionalism? 159
Maps and Stories 162

Part IV: To Know the World

8 **Improvisational Excellence** 169
Improvisation and Environmental Learning 169
An Improvisational Life 171
Language and Place 174
Improvisation and Observation 178
Pattern-Based Environmental Change 180
Random Process 183
Adaptation and Improvisation 186
Improvisation as Flow 189
Networked Improvisation 191
Improvisation, Conversation, and Democracy 193
The *I Ching* as Improvisational Knowledge System 195
An Environmental Change *I Ching* 199
Does Improvisation Emulate the Biosphere? 201

9 **Perceptual Reciprocity** 205
A Glimpse through the Fog 205
The Entropic Mind 208
Floating Bubbles 212

      Becoming a Good Creature   214
      The Great Connectors   217
      Worlds within Worlds   220
      Memory Unfolds in Places   223
      Perceptual Reciprocity in the Neighborhood   225
      Generosity and Environmental Learning   227

A Field Guide to Recommended Readings   231
Words That Matter: A Glossary   239
Environmental Learning Templates   243
Notes   247
Bibliography   257
Index   269

Acknowledgments

Where would we be without our friends, peers, and colleagues? My personal and professional life is, above all, a shared experience. Resonating insights reverberate through all dimensions of a life. Where do insights come from? What prompts inspiration? A kind word, an open ear, an evocative passage, a ray of sunshine, a soft wind on a cloudy day. I perceive myself as one being in an intricate network of relations, aspiring to learn, attracted to people who are striving to know the world, promote environmental learning, and contribute to community well-being. These networks, both visible and unseen, provide support for the book-writing process. The ideas in *To Know the World* are collective interpretations, massaged through many minds, and cultivated over many years. I couldn't possibly thank all the people who offered support, encouragement, and reciprocity, in ways both direct and intangible. Our friends and relations plant so many seeds in the course of a lifetime, and you never know when they will sprout.

I've shared various sections of this manuscript with the following people, all of whom provided kind encouragement and caring critique. Thanks to Simmons Buntin, Alison Deming, Andrea Hedley, Stephanie Kaza, David Lukas, Kathleen Dean Moore, Eric Poettschacher, Christian Rappich, Scott Slovic, Mary Evelyn Tucker, and Tyler Volk.

My good friend, the late Robert D. Kahn, supplied much encouragement along the way.

Thanks to Marina Alberti, Jennifer Bender, Michel Boudrias, Claudia Frere-Anderson, Heather Henriksen, Mary Leou, Joan Raichlin, Grace Wang, and Thomas Webler for allowing me to present ideas to their classes, organizations, and/or staff.

During the writing process, I taught graduate students in the Environmental Education and Communication master's program at Royal Roads University, Urban Environmental Education master's program at Antioch

University Seattle, and undergraduates at Northeastern University. These students worked with the material in its earlier forms. Their discussions and responses were invaluable.

Howard Drossman and Margot Kelley read the entire manuscript, and then sent me comprehensive comments, all of which were pertinent and constructive. They were amazingly helpful. Additional thanks to Margot for suggesting the title.

The MIT Press reviewers provided excellent comments. I've done my best to incorporate changes that respond to their fine suggestions.

A thousand thanks to Beth Clevenger at the MIT Press. Her support, wisdom, patience, and common sense were brilliant. What a pleasure to work with someone who truly understands what you're trying to accomplish!

Cindy Thomashow, my spouse and life partner, listened to every word, often several times in a day, allowing me to interrupt whatever she was doing (within reason), and provided every conceivable means of support, caring critique, and love.

I
Why Environmental Learning Matters

1
The Past and Future of Environmental Learning

Now More Than Ever

In the early decades of the twenty-first century, the mainstream media continuously hyped "breaking news," but the week of September 20, 2019, was truly momentous. Six million climate activists, mainly young people and their adult supporters, from 163 countries, staged over 2,000 events—including mass rallies, school walkouts, and street demonstrations.[1] Although an important focal point was Greta Thunberg's emotional and riveting UN speech, her story was a symbolic narrative encompassing the dreams, aspirations, and fears of millions of global youths from a great diversity of geographic and cultural locations.

That week, a study published in *Science* magazine, "Decline of the North American Avifauna," claimed that since 1970, the number of North American birds declined by 2.9 billion—a drop of 29 percent. The story received widespread mainstream media attention. The *New York Times* (September 25) reported on the most recent Intergovernmental Panel on Climate Change study warning of dire threats to the worlds' oceans. Still on the newsstands, the August issue of *National Geographic*, "World on the Move," was devoted to global migration with the subheading "Seas Rise, Crops Wither, Wars Erupt. Humankind Again Seeks Shelter in Another Place." The September issue, "The Arctic Is Heating Up," included the additional subheading on its cover: "Thawing Tundra Will Speed Up Global Warming."[2]

For people who have been following environmental issues since the 1960s, as I have, these are stories we're accustomed to. Over fifty years later, the direct, tangible impact of climate change, threats to biodiversity, and species extinction are reaching a new threshold—something we surely anticipated. What's different is that the impacts now are so tangible, dramatically impacting the lives of just about everyone who lives on the planet. These events are no longer the supposed special interest of environmental advocates. They are the source of profound human and more-than-human suffering. The young climate activists were born into a world where they worry that their future is limited, their opportunities are diminished, and the earth is in danger. They are fed up with the inaction of global leaders. And they have had enough.

Despite the best efforts of environmental education to deepen awareness of these questions, we have so much more work to do, and much less time to do it. We must embrace the urgency of this ecological planetary emergency. This is a book about environmental learning—how we broaden the scope of our educational efforts, and make connections between the seminal issues of our time, such as migration, race, inequity, climate justice, and democracy—and our understanding of the biosphere. My impression is that climate activists understand these relationships because they live them every day. Our educational work is how to deepen that understanding, use it to inspire widespread action, further clarify the connections, and weave the individual narratives of agency and action into a collective global narrative of constructive change.

I am proposing that we revitalize, revisit, and reinvigorate how we think about our residency on this planet, and am interested in educational approaches for doing so. For all the accomplishments of environmental education (and there have been many), we have not been able to stem the inexorable decline of global ecosystems. Indeed, environmental education is often trivialized, diminished, marginalized, and stereotyped as the province of tree huggers. That's why I use the term "environmental learning" to suggest that our relationship to the biosphere must be front and center in all aspects of our daily experience. This book provides the reasoning, narratives, and approaches for doing exactly that.

We are living in a time of great turmoil, tension, and expectation. It is also a time of hope and promise. I am greatly inspired by the possibilities that await us, young generation of climate activists, and compelling narratives that inspire and motivate them. I wish I could hear all their stories, and weave them together into a tapestry of constructive change, improvisational excellence, and brilliant environmental learning. Our challenge as educators and citizens alike is to tap into that energy and allow it to flourish. *To Know the World* is for the courageous six million activists, the people who support them, and those who are wondering how they should.

What's Next and Why

We have a great deal of ground to cover. Environmental learning encompasses countless themes and the fascinating complexities of their intricate interconnections. Where do we begin and how do we proceed? I've spent fifty years in a variety of educational capacities thinking about that question. Here's what I've learned. There is no one path or sequence for study and learning. You start with what interests you and see where it leads. The role of the teacher is to guide that journey, curate possibilities, ensure that minds stay open and awake, nourish confidence in creative learning processes, and remind learners that education is always reciprocal—giving and taking, remembering and forgetting, practicing and improvising, handing down and passing on, acting and reflecting.

In *To Know the World*, I cover what I think is crucial to the future of environmental learning. The contents reflect my path and interests as well as my experiences as a teacher, my take on the conversations I hear, the people I speak with and learn from, the information I gather, and my interpretation of what all that means. Before I tell you what I intend to cover, I'd like to introduce my own story of how I entered environmental studies way back in the 1960s. As an educator, my strategy is that by clarifying my path, you are inspired to do the same. What is your story, how do you share it, and how will it grow and develop?

What follows is a brief trip back in time when I was the same age as many of the climate youth activists. I believe there is a lineage that connects the past and future of environmental learning. The purpose of this chapter is to amplify that connectivity and contemplate where it leads. I'll start by reviewing the last fifty years of environmental studies. Then I'll explain how the field has changed, and how those changes inform where we go next—what this book covers and why. I will describe the educational perspectives that inform this journey. I have suggestions for how to read this book, make it your own, and use it to maximize environmental learning in your school, neighborhood, and community. How do we expand and enhance multiple narratives of environmental learning?

The Whole Earth Beckons

There's something happening here. What it is ain't exactly clear.
—Buffalo Springfield

I was a college student at New York University in the late 1960s. It was an exceptional time for an inquiring mind and a watershed era in global history. Four movements converged—civil rights, women's rights, peace, and environment. Although these movements had long traditions, the historical matrix of the 1960s allowed them to develop and flourish. It was a time of cultural flourishing too, as the worlds of music and art, inspired by social upheavals, created alternative venues, styles, and perspectives, all linked to new formulations and possibilities for personal identity. The political and cultural reverberations of these movements are just as vibrant and controversial many decades later. They are also the source of much of the so-called values polarizations that are sweeping the globe.

Of course there was no internet then. To understand many of the exciting new ideas, I would scan the bookstores and magazine stands to keep up with the intellectual fervent. On Friday afternoons, I took the subway from the Bronx to Greenwich Village. I typically visited the Eighth Street bookstore and spent hours roaming the shelves. I loved poring over the record bins at music shops, or searching for new and unusual magazines like *Ramparts* and *Crawdaddy*.[3] Many evenings, a group of friends and I would go to the Cafe Au Go Go, where new bands from England or the West Coast—such as Cream, Moby Grape, Procol Harum, and the Byrds—were making their first New York appearances.

One afternoon, I noticed a new and unusual book on the shelves of the bookstore. The front cover featured a photograph of the earth taken from space. The name of the book grabbed me—the *Whole Earth Catalog*. I took it off the shelf and flipped through the pages. It had an atypical layout, with multiple corners, columns, and illustrations. I had never seen anything quite like it. It was divided into compelling sounding sections like "Whole Systems," "Learning," "Shelter," "Community," and "Natural History."[4] In retrospect, this book inspired me to organize my thinking. It provided a way to tie all my interests together, blend the intellectual and practical, and think more deeply about what we now call sustainability. Indeed, as an eighteen-year-old, the *Whole Earth Catalog* represented the future of my environmental learning.

A Book Becomes a Field

Stay hungry. Stay foolish.
—*Whole Earth Catalog*

I pored through the *Whole Earth Catalog,* and took note of the assorted books and "tools" it recommended. It was my first organized environmental curriculum. There wasn't a single course in the entire catalog of New York University's uptown campus that explored ecology or the environment. The first Earth Day (April 22, 1970) preceded the academic field of environmental studies. You could study scientific ecology or forestry, but there were as yet no environmental degree programs. The *Whole Earth Catalog* promised a new interdisciplinary approach to learning that connected many different traditional fields of knowledge.[5]

Within a decade that began to change. I was a member of an expanding cohort of thinkers and learners who created the field that they longed to study. Over the next few decades, there were numerous pioneering programs at all educational levels, and by the year 2000, most North American university campuses had environmental undergraduate and graduate programs.

Environmental studies presented an activist orientation to learning—in terms of both a hands-on, field-based approach to curriculum and the urgency of the issues at hand. Environmental educators and activists alike have played a dual role in producing research that calls attention to critical environmental issues, while promoting policy suggestions and solutions. For most of my career I've worked with teams of people to develop new environmental programs at our home institutions as well as national gatherings of scholars, educators, and activists.

The passing decades represent a great deal of intellectual ground, and there is much to assess about the strengths and weaknesses of these educational efforts. In retrospect, the field has made great strides in promoting public environmental awareness. In the 1980s, concepts like biodiversity and climate change became front-page news. The sustainability idea took off in the early 2000s, influencing universities, businesses, and city governments, becoming a whole new branch of environmental studies. The depths and layers of scholarship, practice, and influence are remarkable. An entire new lexicon emerged. Despite the ever-challenging magnitude of contemporary environmental issues and legitimate concern that environmental values are not sufficiently influential, the public awareness of these issues has reached a new level.[6]

The Emerging Tides of Change

Many years have passed since the first Earth Day. It's not helpful to diminish any of the important accomplishments of the environmental movement. There are waves and cycles of global political trends, however, and environmental concerns do not always fare well. Periods of retrenchment and denial impede policy solutions as well as diminish public awareness. Nor is it sufficient to point out policy accomplishments and feel satisfied. With any significant social change, progress often comes excruciatingly slowly. We try to balance optimism and pessimism. We acknowledge success in order to build confidence, hope, and motivation in the face of difficult challenges.

Feverish tides of change are again sweeping the globe. The same tensions that converged in the 1960s (civil rights, feminism, peace, and environment) inflame as a result of the very environmental issues we've been concerned about—scarcities prompted by natural disasters, political upheavals, and religious extremism; refugees forced to abandon their homes and livelihoods; and interconnected global economic cycles that cause demographic dislocations and migrations, with the attendant impacts on ecosystems and the biosphere.

For the field of environmental studies, in the third decade of the twenty-first century, it's not sufficient to work mainly in the area of conservation and environmental protection. To remain pertinent and responsive, environmental citizens must demonstrate how these convergent challenges are inextricably linked to the fate of the planet. Let's reframe the tides of change as questions:

1. The rapacious exploitation of the biosphere and its life systems continues unabated. How do we best communicate the necessity of ecosystem thinking?
2. There is an increasing disparity between rich and poor. How do we promote economic equity and social justice in cultures of materialism and entitlement?
3. There is great apprehension concerning the integration and separation of global cultures. How do we promote intercultural understanding and cosmopolitan thinking in the midst of nationalist responses and ethnic tribalism?
4. Violence, weaponry, and terror compete with deliberation, diplomacy, and collaboration. How do we settle our differences through community democracy, service, and compromise, in the midst of conflict, autocracy, extreme behavior, and fear?

We are now in a time that very much resembles the 1960s, replete with similar dangers and opportunities. The same discussions of polarization that I experienced in my youth are rampant around the world. The dark shadows of racism, fascism, and religion extremism haunt local and global communities. Yet the calls for equity, inclusion, diversity, and environmental sustainability ring even louder. In an era of mass access to instant information, the cultural shifts morph at lightning speed, and everything is moving quickly. The pace of change is much too fast for some and not nearly fast enough for others.

Environmental Learning and the Tides of Change

The task for environmental learning is to promote a deeper understanding of the "tides of change" while empowering students and citizens to promote human flourishing in local and global communities, both rural and urban. In my career and this book, I make the case that deepening your environmental awareness broadens your understanding of these challenges while promoting citizenship and agency. Our understanding of environmental issues grows through the blending of concepts and practices. Ideally, environmental learning is adaptive, envisioning creative opportunities for study, service, action, and vocation. The world is constantly changing, and so are we. Environmental learning and citizenship must always anticipate the future, and provide constructive solutions. In *To Know the World*, I propose ways to do that.

But before I proceed, I'd like to explain and justify why environmental learning matters, why it's the river through which all the tides of change flow, and why now more than ever, it must be at the forefront of public conversation. Any controversial public issue—affordable housing and transportation, access to education, health care, immigration, and global trade—always has an ecosystem context. The availability, distribution, utilization, and recycling of ecosystem services is the context for prosperity as well as scarcity. That awareness isn't always evident and it must be. There can be no long-term strategic solutions for community well-being at any scale without consideration of ecosystem services.[7]

As people spend increasing amounts of time using screens, driving in cars, riding in airplanes, and working indoors, their experience of the natural world is mediated, diluted, and filtered, thus diminishing sensory and sensual awareness of the biosphere. It often seems that technology enhances perception, and sometimes it does, but the diminished frequency of direct exposure to the biosphere limits our visceral experience of natural systems. Cultivating deeper awareness of our ecological place enables us to come out of our heads and into the biosphere.

The overwhelming presence of humans on the planet along with their ubiquitous fascination with themselves and obsession with consumerism breeds a collective narcissism. The best antidote to this human-centered engagement is to study and observe more than human nature—the birds and bees, if you will. You can't understand the meaning of humanity if humans are the only species you care to interact with. Whether it's the menagerie of microbes that live in your body, the trees in your local park, the birds that arrive at your feeder, or the river that flows through the heart of your city, observing the natural world is the last chance you have to understand what it means to be human on a planet with approximately 8.7 million species.

An ecological perspective on the origins of humanity suggests that we are a migratory species, roaming the planet for suitable habitat, interacting with other species through competition for survival or as coevolutionary partners, diversifying as environmental conditions change, and trading with other roaming bands of tribes. Trace your ancestry merely four generations, and you'll find a genetic soup of mixtures and hybrids. Race and ethnicity are temporary cultural constructions. We are one interconnected species, late arrivals in a complex evolutionary system, bound together as a global phenomenon.

Humans are endowed (through evolution) with the capacity for social learning, reflective awareness, and creativity. We have the potential to appreciate beauty and grace, conceptualize the vast reach of space and time, experience reverence, and contemplate questions of meaning and purpose. That inheritance requires reciprocation, taking the form of stewardship, and fulfilling that potential in service of both humanity and the biosphere.

I believe that environmental learning builds our capacity to engage with these questions, promotes community-based discussions and solutions, liberates how we think about education, and cultivates virtue and character. This book provides approaches for elaborating on this potential and how to catalyze the change that is already happening.

This is a book about environmental learning, how to bring it to the forefront of education, how to expand its impact, and why it's fundamental to human flourishing in the biosphere. Environmental learning matters now more than ever. How do we enhance its prospects and possibilities? And where is it happening?

A New Generation

There's a new generation of environmental change makers who are having a profound impact on communities around the globe. The emerging environmental movement of the early twenty-first century has a new shape and form. Visit any grassroots community-based environmental project to speak to the staff and participants. You will find young activists who understand the necessity of working in diverse communities to promote constructive change. They are concerned with environmental issues, but they are equally committed to approaches that emphasize diversity, equity, and inclusion. They are flexible users of social media, and utilize these skills to build coalitions and promote their ideas.

If you look at the demographics of new social and community movements, you're looking at engaged millennials, people of color, and a great variety of groups that in one way or another are feeling dispossessed. What do these different generations and diverse sensibilities have in common? What can they learn from each other? How do they merge into a coherent, interpenetrating, responsive, and resilient social movement?

Often people from communities of color are on the front lines. They aren't asking for more wilderness, more hiking trails, or wonderful ecotourism resorts. They want opportunity, agency, participation, and equality. Environmental pollution is inevitably worst in the poorest, least white neighborhoods. Affordable housing, access to public transportation, access to inexpensive and nutritious food, nearby parks and recreation facilities, the ability to stay put in a gentrifying neighborhood, affordable health care, clean water to drink and bathe in, and good schools—these are the environmental challenges for the great majority of people who live on this planet.[8]

Environmental learning must address biodiversity and climate change by also dealing with the consequences of inequality, racial divides, and the dislocation of people and cultures. You can't address these inequities without understanding the necessity of intercultural understanding. Hence biodiversity meets diversity, equity, and inclusion. People concerned with these interconnected issues know that those who suffer most will be those in the most deprived economic circumstances. We know, too, that resorting to violence and oppression is a pathological response to the threat of deprivation. It's not the deprived that we fear. It's the prospect that we may become deprived as well.

The Ecological Knowledge Gap

Simultaneously, the ubiquity of social media and electronic communications is shifting how learning occurs, the context for environmental perception, the fate of higher education, and the future of work. There are dozens of new professions reflecting these trends—including information design, social networking, big data compilation, media curating, social entrepreneurship, game design, public art, and impact investment—and they are attracting aspiring youths who understand how much impact they can have with mastery of these skills.

Twenty-first-century science is unleashing extraordinary capabilities—advanced technological sensory systems, earth system monitoring, artificial intelligence and quantum computing, genomic understanding and manipulation, neurobiology and consciousness, networked science, big data, and algorithmic simulations. These are astounding human accomplishments and provide us with exceptional tools for better understanding the evolutionary context of human actions in the biosphere.

Yet this reflects an ecological knowledge gap. There's a widening gulf between highly trained environmental change scientists who are at the cutting edge of research, and a broader, interested public that is concerned about issues of environment, equity, and democracy, and has practitioner experience and local knowledge, but does not have access to either the substantive research or emerging thought processes that provide insights about environmental change. The gap is even greater among those who have merely a passing interest in environmental concerns. Only a comprehensive emphasis on environmental learning can address this gap.

A new generation of highly trained interdisciplinary environmental scholars is integrating earth systems science, evolutionary ecology, biogeochemistry, network theory, urban planning, and sustainability science with new approaches to social and emotional intelligence, decision-making systems, behavioral change, and concepts of human flourishing. The blending of these fields supplies us with the tools and applications to address seemingly daunting planetary challenges. We must make these approaches more accessible.

At the same time, we are witnessing a resurgence of nationalism, tribalism, and ignorance, often transmitted as fake news and propaganda, reflecting a desire to find simple solutions to complex issues while glorifying an idealized past. This mind-set threatens democracy, open societies, individual autonomy, critical and creative thinking, and the very practice of science. Environmental learning must counter these trends.

Thinking Things Through

The choice is between forcing the description of the world so that it adapts to our intuition, or learning instead to adapt our intuition to what we have discovered about the world.
—Carlo Rovelli, *The Order of Time*

Thinking things through.
Thinking through things.
Things through thinking.
Through thinking things.
—Jim Dodge, *Rain on the River*

What is an appropriate educational response to the perils of ignorance? How might the ecological knowledge gap be addressed? How do we revitalize environmental learning? How do we expand the possibilities for environmental awareness? How should the field of environmental studies respond? How can environmental citizenship be renewed and transformed? These questions are the essence of my inquiry.

I invite you to participate in a collaborative exploration of environmental learning. I'll cover a wide range of conceptual ground, synthesizing and interpreting collective memories, experiences, and scholarship. Environmental learning deals with so many different subjects and fields of inquiry. The possibilities are vast, like a mycelial network spreading throughout the forest, making connections, and linking communities of species.

Learning, teaching, and reading are reciprocal processes. They inform each other and spark meaningful conversations. We find points of agreement and disagreement. We spawn celebrations and controversies. We bathe in phases of certainty and uncertainty. We endure waves of confidence and insecurity. We move between compassion and selfishness, between hope and suffering. We reflect and expound. That is how we come to know the world. And that is how I hope you read this book.

Each essay offers hands-on suggestions for enhancing your learning experiences. Consider them thought experiments that stimulate new perspectives, enhance your observational capacities, and generate improvisational approaches and solutions. They are intended to both spark curricular ideas and dynamic conversations while offering practical ways to participate with the narrative. They are organized around taking walks in your community, getting outdoors and exploring, talking to your neighbors, and recording the explorations through mapmaking, art projects, memoirs, narrative interviews, or other forms of documentation, thereby inspiring entire portfolios of additional thought experiments.

In *To Know the World*, my aim is to inspire creative thinking about the future of environmental learning. I encourage readers to curate their experiences and take them where they lead. Along the way, I offer my synthesis of the challenging environmental concepts that should inform our thinking; I present interpretations of why those concepts matter, and how they can be expanded. In that way, I hope that my writing and your experiences inform each other.

This proceeds along two connected paths. Each essay is organized around a "developmental interlude": a personal experience I've had that informs the chapter. This is the best way I know to share what I deeply care about. If this seems overly personal, that's because it is. I have great passion for environmental learning. I want to share that passion with you. I also want to share the dilemmas, discontinuities, and anxieties that emerge from that passion. Environmental learning is most engaging as a process of discovery. It should be a celebration of life in the biosphere, but it is not always pleasant, comfortable, or happy.

The body of the essay subjects are organized around metaphoric themes. These themes (briefly described below) are an attempt to connect the diverse strands of conceptual possibilities. I consider them metaphoric because the terms themselves should generate creative intellectual discussion. For example, before you read the essay on constructive connectivity, consider thinking about what that term means to you. These metaphors are organizational tools. Most important, I encourage you to construct metaphors, syntheses, and interpretations that serve your work and thinking.

I believe that learning is fluid, and the boundaries of expression simultaneously expand and contract. An organizational system is a useful structure, but as a learning tool, we must explore what a plan enfolds and what it excludes. The edge of learning is the interface between structure and change. This book encourages flexibility in moving between synthesis, interpretation, and improvisation. Synthesis illuminates the relationships between ideas. Interpretation is a way to reflect on and speculate about those ideas. Improvisation generates creative expressions leading to another cycle of synthesis and interpretation.

When writing or reading a book, I take great pleasure in assimilating, interpreting, critiquing, and applying the intellectual content. What a gift to contemplate the world of ideas! Ideas matter most when they promote a simultaneous exploration of perception, feeling, identity, mindfulness, compassion, and reciprocity. The most powerful approach for integrating these sensibilities is to cultivate the imagination, balance it with reflection, and manifest it with generosity. Let us balance these qualities of learning.

To Know the World proceeds with a four-part sequence. Part I, "Why Environmental Learning Matters," is an extended introduction, reiterating the urgency of our challenge while illustrating the educational spirit that guides this work. Chapter 1 places *the past and future of environmental learning* into both a historical and personal perspective, makes the case for some new directions in the field, and lays out the plan of the book. Chapter 2 is a reminder that *memory is forever unfolding*. I explore how childhood experiences influenced my learning trajectory, how memory and autobiography generate insight, and how those insights are the foundation of two forms of learning: adaptive and perennial.

Part II, "Environmental Learning in the Anthropocene," investigates the global challenges that inform environmental learning. Chapter 3 goes into more depth regarding the *tides of change* sweeping the globe, and how they set the context for environmental learning and citizenship. Chapter 4 asks whether the *Anthropocene is blowing your mind*, and if the accelerating pace of technological change further separates psyche from biosphere. It suggests that a deliberate focus on ecological place provides balance and perspective in a world of proliferating information and instant access.

Part III, "A New Vision for Environmental Learning," describes three convergent challenges—networks and their manipulation, migrants and refugees, and tribes and territories—why they are the crucial unfolding environmental issues of our times, and how they can be informed by ecological awareness. Chapter 5 is an inquiry into *constructive connectivity*: how networks allow us to create social capital, build bridges between communities, and better understand ecosystem relationships. Chapter 6 suggests that a deeper understanding of *the movement of people and species* is a gateway for considering refugees, migrants, and immigrants, that human migration in the biosphere is a successful adaptive strategy, and why we must allow for compassionate passage through and between habitats, nations, and communities. Chapter 7 introduces the idea of *cosmopolitan bioregionalism*, an approach to the integration of local and global thinking that bridges the divide between urban and rural, insider and outsider, and place based and cosmopolitan.

Part IV, "To Know the World," explores two learning aspirations—improvisational excellence and perceptual reciprocity—that simultaneously promote a deeper awareness of the biosphere and human flourishing while enhancing the prospects for living a meaningful life. Chapter 8 looks at how *improvisational excellence* is the foundation for adaptive flexibility, creative insight, and the ability to recognize and interpret global environmental change, thus enhancing the possibility for making good decisions in conditions of uncertainty. Chapter 9 assumes that there are multiple forms of species and planetary awareness, and human flourishing depends on our ability to tap into that awareness through *perceptual reciprocity*, the ability to learn about biosphere processes and species' behaviors that we may not fully understand, but that embody planetary knowledge and wisdom.

Although I've thought long and hard about the learning sequence of these chapters, there are many ways to read a book. In the spirit of creative improvisation, I encourage readers to find a path that seems most pertinent and appealing. The book is divided into multiple subsections so that you can read a few pages at a time, at a pace of your choosing. I do recommend that you read individual chapters in their sequential order, but there is no reason why you shouldn't follow a unique chapter sequence as a path through the book.

There are four additional reading resources. Please take advantage of the table of contents. The listing of subsections may serve as a map of your intellectual location, a sense of place for the reader, and a way to scan what's ahead and see the whole picture.

I provide a glossary, "Words That Matter," of some of the seminal terms and concepts. The glossary is not a repository of formal definitions but rather a reminder as to the conceptual linkages between ideas. Many of the word-concepts are my attempt to synthesize ideas. I do so by experimenting with combinations, such as cosmopolitan bioregionalism, perceptual reciprocity, and constructive connectivity, among many others. I encourage you to experiment with unique combinations. Who knows what insights you'll generate?

The recommended readings section is another organizational tool. If you want to dig more deeply into the core themes, you'll find some of the books and resources that I find most illuminating. It is both wonderful and daunting that such lists are always expanding. You can never quite cover what you think you need. It's useful to highlight the emerging classics—the perennial books, like good friends, that will always be there for you.

There's also an index of environmental learning templates. These can be used in curricular settings or as mindfulness activities for further contemplating various dimensions of environmental learning. As I stress throughout the book, use them as foundations for your own inventions.

In the next essay, "Memory Forever Unfolding," I share some of my most important learning experiences with you. These experiences shaped my development in profound ways. I hope that while reading them, they evoke many of your own memories. When they do, then you become a participatory reader. Those memories will undoubtedly lead you in different directions. But if we can share them, then we can find common ground—the place where our experiences meet. It's crucial that we share these stories because they are the lifeblood of what matters most in our lives. This book is an invitation for you to share stories of environmental learning. There is no better way for us to know the world.

2
Memory Forever Unfolding

The Branches of Memory

This present moment that lives on to become long ago.
—Gary Snyder, *This Present Moment*

Life has a way of talking to the future. It's called memory.
—Richard Powers, *The Overstory*

Follow me to a small apartment in Fresh Meadows, Queens, New York, during the early 1950s. My world between the ages of three and five was a middle-class, brick housing complex, a typical midcentury, postwar apartment building. I have several dim but residing memories of the playground and my bedroom. I spent hours on the swings, singing songs at the top of my lungs, experiencing freedom as a small child might perceive it. But what I remember most are some of the books I read. I taught myself how to read, and as early as three years old could follow simple books. By the time I was five, I could read just about any children's book. That was my other source of freedom. Now, over six decades later, two of those books stand out. Their deeper meaning was slowly revealed during the course of my educational development. In retrospect, they were fundamental to how I think about learning and my lifelong engagement with environmental issues.

The Golden Guide to the Stars and *Harold and the Purple Crayon* remain with me all these years later. *The Golden Guide* inspired wonder, humility, exploration, and reverence, providing glimpses of infinity along with the vastness of time and space. The *Purple Crayon* was an improvisational guide for adapting to changing circumstances. I will also tell you about the fog man, the truck driver who sprayed DDT in my suburban development, and how I learned about toxicity, sowing the seeds of fear and vulnerability. At an early age I grappled with openness and vigilance. Sixty-five years later, I'm still trying to figure out how to balance these states of mind.

These memoirs are examples of developmental interludes. I'll explain why I think such narratives are a foundation for environmental learning. I'll also distinguish between perennial learning (the educational wisdom of the ages) and adaptive learning (how changing times call for flexible approaches), and how these processes are mutually enhancing, allowing us to generate wisdom as we face the unknown. Wisdom forever unfolds too, and as it does so, it reaches into the furthest corners of our collective experience—past, present, and future. As Lewis Hyde reminds us in his wonderful book *A Primer for Forgetting*, "The truth about who you are lies not at the root of the tree but rather at the tips of the branches, the thousand tips."[1]

The Golden Guide to the Stars

You need no equipment to see and study thousands of stars. This book will point the way to hours of interesting study with nothing more than your two eyes.
—Herbert Zim, *The Golden Guide to the Stars*

One Sunday, my parents took me to the Hayden Planetarium in New York City. I was awestruck by the immensity of the virtual cosmos. I experienced wonder in a palpable way—the night sky, planets, and vast expanse of outer space. I sat in the circular auditorium gazing at the virtual heavens. What a bonanza for a young imagination. Outer space appeared accessible, promising, and truly wonderful. We visited the planetarium gift shop, and my parents allowed me a souvenir. I asked for Zim's *Golden Guide to the Stars*. I savored that book for years to come. I didn't understand all the text in the book. Still, it took me on magnificent journeys throughout the solar system, illustrating comets, imaginary planetary landscapes, moons, stars, and asteroids. I was enamored with the two-page spread depicting a beam of light passing through a prism, resulting in the colorful spectrum. In my mind's eye I can still see many of the book's illustrations. In those early years, I first came to know the world by using my imagination to visit outer space and gain a perspective on the earth as a planetary body. This was my first lesson in scale.

There was a dark side too, revealing the limits of my insufficient capacity for understanding. Periodically I would have terrifying nightmares. I can't recall anything substantive about them. Yet I can still conjure the deeply unsettled feelings. The dreams had to do with inexplicable proportionality. I was outrageously small or large in relationship to my space. In retrospect, I was swimming in a weightless dimensionality, unable to find my bearings, terrified by feeling disembodied, and befuddled by a sensational surreality beyond my experience or any rational explanation that would make sense to a five-year-old. It was my first encounter with the void. The world (and universe) was both wonderful and inexplicably daunting, and perhaps forever unknowable as well.

The Golden Guide to the Stars unleashed my imagination. I knew that my mind (and dreams) could wander far and wide. Books were an inspiration that could launch countless journeys. As a young child, they allowed me to expand the territory of creative exploration. I was delighted with the prospects. To think that a small book could have so much power!

Harold and the Purple Crayon

One evening after thinking it over for some time, Harold decided to go for a walk in the moonlight.
—Crockett Johnson, *Harold and the Purple Crayon*

All explorers need a guide. It's easy to wander off the path, or encounter landscapes, people, animals, or situations that you aren't quite ready for. A five-year-old traveler also needs guides, especially when the journey is propelled by an active imagination. I couldn't assimilate all that I was learning. I had many irrational fears. After reading through *The Golden Guide to the Weather*, I was worried that a harmless cumulus cloud on a warm summer day could become a dangerous thunderstorm. I distrusted tunnels and elevators, unable to see where I was going. I was frightened of speed and loud noises.

I found a book that gave me confidence. It helped me confront my fears by demonstrating that I had the power to construct, coordinate, and curate my expanding imaginative reality. You see, Harold had this magic crayon.

Harold used the crayon to draw the landscape of his evening escapades. At the start of his adventure he required a moon for light, a path to walk on, a tree to explore, and some apples to eat. He drew a dragon to protect the apples, but it was so frightening that the hand holding the crayon began to shake. He inadvertently drew ocean waves and fell into the water. Harold improvised. He drew a boat and took a sail guided by the light of the ubiquitous half-moon. Other adventures follow, including a picnic with a "hungry moose and a deserving porcupine," a climb up a mountain to get a better view, falling off the mountain and constructing a balloon in the midst of his free fall, and finally as he begins to tire, a search for his home and bedroom. He draws a house and then an entire city neighborhood, with many tall buildings and countless windows. But the windows all look the same, and he can't identify his home. As he gets increasingly worried, he has a brilliant insight. He remembers the moon shining in his bedroom window. So if he can draw his bed and the window with the moon shining through, then he can find his way home. He does so and falls off to sleep.

Harold and the Purple Crayon is one of the most adored children's books of the twentieth century. It has sold over two million copies. It has never gone out of print. There are six other *Harold* books, multiple animations, and a thirteen-episode HBO series. There's even curriculum on how to use the book to teach young children philosophy. Harold is not only popular, but the lessons of his journey are relevant for children and adults alike.

Here's why I remember it so vividly. Harold is seeking adventure and discovery. He does so through his fertile imagination. He has no map. His experiences are spontaneous, wonderful, and scary. They lead to predicaments. He regains his balance through improvisational thinking, using the crayon to create structures that emerge from the tribulations of his creativity. He is always learning something new. And then when he is fatigued and overwhelmed, he searches for his home place. He arrives home by drawing an image of what's comfortable and familiar.

I first read *Harold and the Purple Crayon* the year it was published (1955). It was one of my first lessons in how to explore the world. Over the course of my lifetime, the substance of the adventures have changed considerably. Yet the sense of insecurity and risk, waves of confidence and doubt, challenges of uncertainty, uses of creativity and imagination, urge to discover new ideas, necessity of rethinking what you thought you knew, inevitability of mistakes, and ability to flow with the circumstances—these are not only the challenges of a lifetime but as you'll see throughout this book, the foundations of environmental learning too.

Harold's culminating lesson is the celebration of a secure home place. He desires mobility. With mobility he encounters adventures and unpredictability, but he also knows that he can locate and reconstruct his way home. Home place (however it is defined) is a conceptual ballast that supports transience, impermanence, exploration, and discovery, and allows the creative imagination to flourish.

Here Comes the Fog Man

Beyond the light there are many kinds of darkness at the edge of knowledge.
—Caspar Henderson, *A New Map of Wonders*

The branches of memory may reveal dark shadows—disturbing images that deeply impacted how you came to know the world. One memory starts as a frolic and ends as a headache. In 1956, we moved to a suburban community on the south shore of Long Island. Our development was essentially built on a wetland that drained into a series of beaches bordering the Atlantic Ocean. During the summer months there were plenty of mosquitoes. The solution? Spray DDT! A man would come in a small jeep and dispense a cloud of chemicals resembling a misty fog. We called him, as I mentioned earlier, the fog man. I remember running through "the fog," delighted that I had access to the same clouds I observed in the sky. I soon realized that these fog adventures would typically result in a headache, and from then on I stayed away. In my adult years, as a thirty-nine-year-old and then more recently, I've had cancerous but encapsulated tumors removed from my body. I often wonder whether the fog man seeded those mutations.

Around the same time, the United States and Soviet Union were engaged in a feverish and insane experiment of aboveground nuclear testing. You could watch images of mushroom clouds while perusing the evening news. In school, we would have drills to protect us in case of air raids. The teacher instructed us to put our heads down on our desks. I was only six years old. But it wasn't hard to figure out that if the Russians dropped an A-bomb on New York City, this was feeble protection.

What lessons emerged? First, I experienced cognitive dissonance, or the awareness of conflicting possibilities. My mind was twisted by the prospect of a misty fog leading to poisonous headaches. How could the expectation of delight prove to be delusional? I longed to explore the clouds that were drifting on the ground. But I didn't realize they were a human-made gas, linked to a toxic chemical that would soon be banned. This never dampened my enthusiasm for investigating landscapes and weather, but it did teach me about prudent vigilance. This theme reemerged in my professional work decades later as a crucial existential and educational challenge: how to engage a sense of wonder at exploring the biosphere when you simultaneously learn about threats to biodiversity, climate change, and species extinctions.

Second, I learned that authorities were perfectly capable of misleading people. Did my teacher really think that putting my head down on the desk would protect me from a nuclear explosion? Why did we have such weapons when their use sparked mutually assured destruction? I didn't ask such specific questions until later in childhood, but my sense of fear was surely enhanced. And more important, I began to question the wisdom of adult decisions.

Knowledge, I learned, was a double-edged sword. It could fuel my imagination and scare the living daylights out of me. This mirrored a dual message of the 1950s—my first exposure to a lifelong conceptual incongruity. Check out appliance, television, or automobile advertisements of the 1950s. They promise the glorious prospects of unbounded affluence, showcasing shiny and sparkly consumer goods. Yet newspaper headlines and newscasts discussed the Cold War while government officials recommended that ordinary citizens build bomb shelters. Of course it's easy to dramatize this in retrospect, but these memories remain vivid, and they informed my hopes and fears, my seductions and aspirations, and the images of success and failure that are still prevalent six decades later.

Developmental Interludes

The entire evolution of science would suggest that the best grammar for thinking about the world is that of change, not of permanence. Not of being, but of becoming.
—Carlo Rovelli, *The Order of Time*

I'll conclude this brief snapshot of vivid learning memories for now, although I could easily cite dozens of additional incidents. Some of those reflections will emerge later in this book. I call these passages developmental interludes. They are dynamic approaches to life cycle learning. It's instructive, regardless of your age, to trace how you learn through a lifetime, consider what's changed, track the integrating themes that continue through the stages of your life cycle, and observe that the essence of how you learn is both perennial and adaptive. I use the term "developmental" to stress that at different stages of conceptual, cognitive, and emotional maturity, we have perspectives that reflect the time, age, and place of our situation.

I believe there are recurrent themes over a lifetime—interests, ideas, dilemmas, and ways of perceiving the world that form your personal narrative. If you take the time to meditate and reflect on different stages of your own development, these narrative themes become clearer. You gain perspective on how your actions have a broader context and recreate themselves as learning opportunities, and how your life is, in fact, a cycle of possibilities, unfolding in dynamic and interesting ways, offering unique insights and wisdom.

Of course, your body and mind change with each breath and every thought moment. You recycle cells and ideas alike. There are narratives, however, that tell the story of your life, integrating coherent themes by coordinating and interpreting the process of developmental change. In this book, I am asking you to look carefully at those themes. They are a pathway toward a better understanding of the learning habits of a lifetime. As you find deeper meaning in those memories and reflections, you will better understand how you come to know the world, in a way that is personal and meaningful.

I will suggest, too, that revealing these narratives should be a collaborative effort. What are the ways our shared experiences are similar and different? This is an educational plea for the necessity of both multigenerational and multicultural exposure. You can't know the world just by knowing yourself. Indeed, you come to know yourself by knowing others, and that's when deep learning happens. That is the essence of lifelong learning. By capturing my own developmental interludes, I hope to trigger some of yours.

This reflects my hands-on educational philosophy as well as the writing strategy of this book. The most interesting learning opportunities unfold from your personal experiences, especially when they are shared with others. When your personal narrative is curated with purpose, and then melded with ecological observations, synthesized through interpretation, and organized by analysis and theory, your learning expands and your capabilities flourish. Developmental interludes help you expand your sense of self. They prepare you to engage in profound environmental learning.

Perennial Learning

> Wisdom begins with awareness, of the self and the world outside the self; it deepens with our awareness of the inherent tension between the inner "I" and the outer world.
> —Stephen S. Hall, *Wisdom*

Harold and the Purple Crayon is just as meaningful today as it was in 1955. Yet we also should be aware that in these times, Harold is as likely to take his journey on an app or video. Harold's adventures are perennial because they tap into challenges that face every generation, regardless of the medium.

The human imagination finds many ways to travel through space and time, and experiment with scale, perception, and perspective. Artists, visionaries, mystics, and scientists use the imagination to experiment with scale, expand awareness, and explore and discover the world. Harold just happened to use a purple crayon. It could have been a musical instrument, the written word, a form of dance, a microscope, a telescope, or a sextant. These are educational mediums. Whether they are used for pleasure, performance, measurement, or experimentation, they tap into the creative capacity, stirring the imagination, generating insight, and stimulating emotional responses.

You may not be conceptually ready for what you encounter. And you may not be able to assimilate what you learn. Memory and the imagination reverberate throughout your awareness, allowing you to construct multiple realities, playing out dynamic scenarios that may never come to pass. The imagination can visit realms that are both wonderful and terrifying. That's why we require guidance, collaboration, and mentorship, and why learning is a multigenerational activity in which we pass ideas from one time period to the next. This happens by virtue of the stories we tell each other, whether around a campfire, at a family gathering, in a classroom, or through a magnificent book. Learning is perennial when it taps into wisdom that penetrates many generations. It's likely that the dilemmas you face, the questions you have, and your hopes and aspirations connect you to other people and cultures, both through the ages and over the course of your lifetime.

Perennial learning is awareness of the wisdom of the ages. Yet the context and milieu of environmental learning is always changing. For example, promising new technologies easily take over our lives, and in so doing, alter what we observe and how we perceive. Our social relationships are powerful forces that expose us to waves of trends, styles, and implicit conformities. How do we find a path through the changing flows and circumstances of our social as well as technological encounters?

Adaptive Learning

The whole world's living in a digital dream
It's not really there
It's all on the screen
Makes me forget who I am
I'm an analog man
—Joe Walsh, "Analog Man"

The latest research in evolutionary biology and cognitive neuroscience is complex, emergent, and often controversial. A tentative consensus, however, suggests there is a somatic and evolutionary substrate that informs sensory perception, structures our imaginative capacities, and enables us to form concepts and ideas. Exciting new research examines the remarkable plasticity of the brain in relationship to cultural evolution. In *Evolution in Four Dimensions*, Eva Jablonski and Marion J. Lamb propose an evolutionary synthesis that describes four inheritance systems: genetic, epigenetic, behavioral, and symbolic. That means there are many pathways for transmitting information (and hence knowledge) between and within generations. Kevin Laland's work as outlined in *Darwin's Unfinished Symphony* indicates that the most important contributing factor to the rise of human culture is our capacity for social learning. What happens when you consider the educational implications of this proliferating research? Surely it means we must pay attention to understanding the adaptive qualities of active minds.

I know that my granddaughter finds Harold's purple crayon as engaging as I did. And just as I used to love to run on the playground, I take pleasure in admiring her great joy at kinetic learning—climbing on whatever surfaces she can while challenging me to races I can no longer come close to winning. I also marvel at her ability to sit with a computer tablet and make her way around the interface entirely on her own. Just as I learned to read with minimal supervision, she can navigate computers through trial and error by observing others as well as just figuring it out.

My grandparents grew up in an era when there were no airplanes (invented in 1903) and just a few automobiles (invented in 1885). My granddaughter watches videos on a tablet. Born in 1950, I stand in the middle of four generations, a child of the mid-twentieth century, a transitional era, that predates the so-called digital age. Like Walsh, I'm an analog man, but I am intrigued and thrilled by the learning possibilities of the digital realm. Yet I know that despite my familiarity with computers and the internet, and all the ways I am deeply enmeshed in the digital age, my exposure came well into adulthood. My granddaughter has always known tablets and the internet, and that changes everything.

I do not intend to levy judgment. I prefer to reveal the depth of my unknowing of the educational implications of how quickly learning environments change. The technology proceeds apace. We adapt and learn together. As Jablonski, Lamb, and Laland, among many other scholars, point out, there are many vectors of information transmission and social learning, and we cannot pretend to understand the various ways they unfold.

Therefore it is undoubtedly presumptuous to write a book that covers the future of environmental learning. The ecological and evolutionary dynamics of rapid global environmental change are complicated enough. Superimpose the rapidity of cultural shifts, transmission of information, and perceptual and sensory impact of advanced technology—the interpenetration of memes, dreams, and machines—and we realize how much there is to absorb.

Our challenge is to consider the necessity of adaptive learning along with the ability to assimilate, interpret, and synthesize the rapidly changing circumstances of our planetary condition. You work with what you have while attempting to anticipate what comes next. You learn to improvise, confront uncertainty, cultivate open-mindedness, and listen well, and sprinkle this formula with a healthy dose of humility.

And then you look up at the sky and see the same constellations your ancestors did. Or you take a walk in the woods, and marvel that the lichen growing on the trees and rocks have an ancient lineage stretching into the deep time of the earth's history, acknowledging that some Arctic lichen live for eighty-six hundred years. Or you feel awe that the atoms in your body were formed when a distant star exploded in a past so remote you can barely conceive it.

Yes, it's necessary to consider how the changing milieu of contemporary life requires adaptive considerations in educational settings. But remember that an ancient work like the *I Ching* is also a "book of changes," and its timeless advice reflects on the inevitability of change in all human endeavors.

Depending on your age, culture, background, values, and interests, there are many paths to environmental learning. This includes varieties of developmental sequences, cognitive abilities, and multiple intelligences. Let's keep this in mind as we proceed. Your path is an educational narrative—one strand in a collective narrative that yields more generalized insights. The best educational approaches weave individual and collective experience, while calling attention to what's perennial and adaptive.

An Ocean Voyage

Hexagram 26, Great Restraint
Firmness and Strength,
Substance and Brilliance
Daily renewal
Of Inner Strength.
A Firm Line is in Top Place,
The worthy are honored,
Strength is contained,
This is Great Truth.
Not eating at home
Is Auspicious,
It nurtures the worthy.
Crossing the Stream
Resonates with Heaven.
—John Minford, *I Ching*

When I was twenty-three years old (1973), like many young people that age, I had wanderlust. I had just finished a master's program in history in which I spent countless hours reading and writing. My mind was surely challenged, but I felt hemmed in and claustrophobic, not wanting to spend my twenties in libraries. Through a series of fortunate inquiries and connections, I secured passage on a coal freighter traveling from Norfolk, Virginia, to Taranto, Italy. This opportunity thrilled me. Tired from my cloistered experience, I brought no reading material with me, only my guitar and pack.

On arriving at the ship, I met the captain. I was under the impression I would be put to work, but he said that wasn't necessary. This was a ten-day voyage. I realized there would be nothing for me to do except play the guitar and gaze at the ocean. I had a few hours before the freighter departed. I hitched a ride to downtown Norfolk searching for a bookstore. All I could find was a pornography bookshop, but in the back it had a small shelf of literary books. I grabbed *The Glass Bead Game* by Hermann Hesse and John Blofeld's sparse translation of the *I Ching*.[2]

During the first evening, I sat in my cabin and went through the *I Ching*'s coin-throwing procedure. I asked whether this voyage was the right path for this phase of my life. My hexagram was number twenty-six, translated by Blofeld as "The Great Nourisher." The first lines read, "The Great Nourisher favors righteous persistence. Good fortune results from not eating at home. It is a favorable time for crossing the great river (sea)."[3] This was one of the first great independent adventures (voyages) of my life. The seemingly random process of selecting a hexagram was incredibly reaffirming.

Thirty-three years later, I was offered the presidency of Unity College in Maine. Taking on the leadership of a small university would be another provocative adventure. I've always saved the *I Ching* for new challenges, and only use the random process when I want to shake up my otherwise-rational decision-making approach. As I was preparing for what would be a major life change, I once again threw the coins to select a hexagram. I used a different, more comprehensive translation, this one by Jack Balkin, a constitutional law professor at Yale University. His approach is geared toward issues of leadership, as are many of the *I Ching* translations. Remarkably, I arrived at the same hexagram, named in this translation "Great Accumulation."

The "Judgment" section of the text reads,

> Cultivating a determined and steadfast character is essential, because if you wish to remain at the height of your powers you will need to hone your skills continually and renew yourself daily. . . . [G]ood fortune will come from working for the public good rather than for private advantage. Devote yourself to the perfection of your talents and to the achievement of something beyond your narrow self-interest, and you will have both the power and the vision to achieve great things, symbolized by crossing the great river.[4]

I have a modest collection of *I Ching* translations. And my knowledge of the work is at best equally modest. It was written many, many generations ago, sifted through Confucian, Taoist, and Buddhist minds, and then filtered through Western scholars from different intellectual traditions. The yin and yang lines yield archetypal nature-derived trigrams that recombine as sixty-four hexagrams, interpreted through observations of the natural world, human behavior, and the nuances of governance and leadership. It is arguably one of the most enduring, intriguing, and mysterious books to emerge from human culture. When I study the book, I feel as if I'm accessing an intellectual mixture of the world's great wisdom traditions, organized around a primordial, archetype knowledge system.

The ten-day voyage was slow and steady. For virtually the entire trip, there were no human signs other than the daily doings of the crew. I played my guitar, studied those two books, and gazed at the ocean. For the first nine days, I contemplated sea and sky, remarking on the absence of human signatures. A half day out from the Strait of Gibraltar, ships appeared. The closer we got, the more ships I saw, as if they were all pulled by a magnet through a funnel. Finally, the sea traffic was dynamic. There were hundreds of boats of all conceivable shapes and sizes, carrying the world's cargo through a narrow passage into the Mediterranean Sea. I wondered what it might have been like only five centuries earlier to sail in the opposite direction, westward through the passage into what was then the great unknown, to be a passenger on such perilous journeys. I wondered, too, about the great contrast of sailing across a wild ocean, and then entering the busy world of human commerce and exchange.

Writing a book is a challenging voyage in its own right. It's not as physically perilous as crossing an uncharted ocean. But it's a powerful learning experience. And there are times when you confront the great unknown. I write a book when I feel that I need to dig deep within myself to pull out what is at the edge of my awareness. It's an exploration of inner space made public. You spend thousands of hours devoted to exploring ideas and experiences that you hope will be of wider interest. Despite your best plans and whatever structure you have in mind, there will be meandering paths, looping themes, and reiterative cycles of insight and frustration. You meld perennial learning (what it is you've always known) and adaptive learning (how to apply that knowledge to changing circumstances). You develop rational, sequential approaches while looking for creative ways to express them. You improvise as necessary while always returning home to the seminal questions that inform the work. Still, your ideas congeal and then decompose, reforming throughout the course of a lifetime. The learning process, like your memory, is forever unfolding.

So grab your purple crayon and use it well. Take the ideas in this book and carry them to new places. Use them to expand your thinking and challenge yourself to challenge others. Delve into your own memories and experiences to allow the theoretical material to come alive. Stay home and observe the natural world. Use your animal senses. Discover your roots by learning about what's right in front of you. Then you can stretch far and wide. Cross the stream so you can resonate with heaven.

II
Environmental Learning in the Anthropocene

3
The Tides of Change

Pacific Northwest Change Makers

Between 2016 and 2017, I was living in Seattle, Washington, serving as a Sustainability Catalyst Fellow at Philanthropy Northwest. My assignment was to profile community-based sustainability projects that emphasize diversity, equity, and inclusion as intrinsic to their mission. I spent several months interviewing community activists, hoping to better understand their goals, approaches, and challenges. I visited projects in both urban and rural regions, in a variety of ecological landscapes. I took road trips to coastal Alaska, the Crow Nation, rural Montana, and the Puyallup Watershed that flows from Mount Rainier to Puget Sound. I visited small cities like Missoula, Montana, and Walla Walla, Washington. And I learned about projects in Portland, Oregon, and Seattle. My final report, *Pacific Northwest Changemakers*, profiles eight projects, all with uniquely interesting stories.[1]

Despite the differences in geography and emphasis, these projects share many similar themes. First, they are place based, grass roots, multigenerational, and cross-cultural, while aspiring to balance economic prosperity and ecological integrity. Second, they view environmental concerns as intricately connected to community well-being, tackling issues that matter most to community members even as they stress entrepreneurship, consensus building, and inclusive collaboration. Third, project staff include a youthful blend of activists representing a dynamic cohort of emerging leadership. During my travels, I was inspired by these staff members, impressed by their commitment, encouraged by their community orientation, and excited by their desire to set up roots in the place where they work. I learned a great deal by listening to their stories and gained a better understanding of how they serve as agents of change.

Along coastal Alaska, I spoke with Klawoch, Tlingit, and Haida communities that are redefining sustainability to encompass tourism, forestry, salmon fishing, environmental entrepreneurship, and indigenous sovereignty regarding ecosystem services. These indigenous people work closely with their Caucasian colleagues, many of whom were born in Alaska and have returned to work in the communities where they were raised. In almost every case, these "change makers" left Alaska to get an education, join the military, or broaden their experience, and then returned home to apply what they learned, melding their skills with the spirit of place so deeply entrenched in their awareness. In Seattle, Tacoma, and Portland, I spoke with people from a variety of cultural backgrounds who have clearly stated environmental values, but understand they must find ways to broaden the meaning of environmental quality so it also includes affordable housing, access to health care, educational opportunities, and community rootedness. They prioritize working with immigrant populations that risk displacement in a booming economy. In rural Montana, I met with ranchers who understand their way of life will only survive if they find ways to integrate conservation stewardship with economic opportunity. In Missoula, I spent a day with a food justice program that uses urban gardens to provide opportunities for homeless teenagers, elders, and the unemployed. A project at the Crow Nation in Montana is emphasizing food sovereignty, traditional ecological knowledge, and microentrepreneurship to supply employment opportunities, especially for young people.

These projects are investing in local communities, building commitment to place, and aspiring to include multiple stakeholders. They understand the necessity of long-term involvement, cultivating trust and capacity over time, while listening to the various needs of community members.

I visited these projects shortly after the 2016 election. Like many people, I was dismayed at the polarization in Washington, DC, toxic news cycles, overriding emphasis on the president and his discontents, and seeming inability of breaking through the gridlock. Above all, I was tired of the recurring narratives of doom and incapacity.

There's another narrative—a much more hopeful one that paints a different picture. I visited just a small sample of change-maker projects. I could have easily chosen several hundred more in the Pacific Northwest. There are thousands of projects like these all over North America, and tens of thousands scattered around the globe. Yet if you follow the nightly news, you would never know about them. There's an alternative narrative. It presents a constructive, hopeful vision of a sustainable environmental future. I don't wish to create a false impression. All these projects face difficult challenges—from raising sufficient revenue to building trust across diverse communities. There are always skeptical critics. And many rely on ephemeral funding cycles from foundations or the government. There's no guarantee that their various approaches to community-building entrepreneurship will succeed. Still, many of these projects are already several decades old and deeply embedded in their communities, garnering respect and influencing public policy.

Let's briefly take a closer look at how the tides of change impact the Pacific Northwest, illuminating why these change-maker communities are relevant both locally and globally, and why they will remain so for years to come.

The Pacific Northwest: A Brief Profile

The Pacific Northwest of the United States enjoys spectacular beauty, abundant natural resources, and rich ecological and biological diversity. A regional map shows twenty-seven separate bioclimates, stretching from Alaska to southern Oregon to Wyoming, including coastal rain forests, deserts, fertile lowlands, and interior mountains, each featuring distinctive subdivisions. There are dozens of indigenous tribes with unique languages and cultures. There are immigrant groups from a hundred countries. The region includes one of the fastest-growing cities in the United States (Seattle), a bustling transportation corridor (Interstate 5), and hundreds of rural communities.

The history of Pacific Northwest settlement is a story of natural resource extraction, especially in timber, mining, agriculture, livestock, and fisheries. Indigenous peoples frequently practiced what we today call "environmental sustainability." Their cultural traditions emphasized multigenerational resource use. In only a century, US westward expansion brought a different approach to the land, resulting in dramatic increases in clear-cut logging, landscape-scale mining, industrial agriculture, and overgrazing as well as the depletion of salmon runs. This process still has a great impact on communities throughout the Pacific Northwest, prompting difficult conversations about conservation, stewardship, prosperity, autonomy, and opportunity.

In many communities, the profits from natural resource extraction yield temporary prosperity. When the resource dwindles, some communities are left behind, with few economic alternatives. This results in unstable boom-and-bust cycles, undercutting the fabric of social life and requiring communities to reassess their local economies.

The impact of boom-and-bust cycles takes various forms in urban and rural areas, but there are similar patterns. Seattle and Portland are examples of boom cities spurred by the arrival of tech and biomedical industries along with an infusion of social entrepreneurs. Some wish to settle permanently. For others, these are way stations in the global economy. One consequence of these intentional migrations is a spectacular increase in real estate speculation; construction cranes are ubiquitous on the Seattle and Portland skylines.[2]

As gentrification spreads, affordable housing becomes a distant memory. The displaced residents, many in lower economic brackets, can no longer afford to live in their neighborhoods. Hence affordable housing, accessible transportation, access to health care and education, and many other basic aspects of community life are under threat. Prosperity also yields displacement. Cities are in the midst of rapid demographic and economic change, and this has ecological, social, and cultural consequences.

Rural communities are on the front lines of natural resource extraction cycles. Conservation of natural resources is crucial to their future. Their challenge is to develop an ethic of stewardship so that the natural resource base generates prosperity while creating alternative revenue options. As the cities boom, many rural areas become backwaters as their young people no longer have ways to earn a living. No place is exempt from the waves and cycles of the global economy. Urban and rural regions alike deal with the consequences for community life.

Although regions throughout North America and the world have vastly different ecological and cultural settings, the dynamic tensions of the Pacific Northwest are familiar: boom-and-bust cycles, threats to environmental sustainability, migrations and displacements, inequitable prosperity rewards, tensions between long-term residents and newcomers, stresses between urban and rural communities, and questions about who makes decisions for a community as well as how those decisions are subject to trends beyond local control. And all these challenges will be further strained by rapid environmental changes, especially those that are climate related.

These are the global challenges (in local and regional settings) that must inform environmental learning. In this chapter, I'll reframe these challenges as emblematic of the tides of change I referred to in chapter 1, labeled as environmental sustainability (ecology), the mixing and separation of cultures (diversity), the distribution of wealth (equity), and the threats to democratic decision making (inclusion). I hope to show that these challenges are not unique to our time, although their impacts are greatly accelerated. Rather, they are intrinsic to the human condition and reflect the evolutionary circumstances of our species. I view them through an educational lens. Is there a learning process that can help us better understand these dilemmas and promote constructive ways to think about them? Can we personalize these challenges while also understanding the big picture?

Two sections provide some necessary background. First, I will review the concept of the Anthropocene, how it represents the culmination of 150 years of environmental thinking, and why it's an essential framework for thinking about the tides of change. Second, I'll explain the global challenge of dislocation and displacement, and why it's the emerging and culminating issue of the next several decades.

Now the Anthropocene

The Anthropocene has reversed the temporal order of modernity: those at the margins are now the first to experience the future that awaits all of us; it is they who confront most directly what Thoreau called "vast, Titanic, inhuman nature."
—Amitav Ghosh, *The Great Derangement*

Forget for a moment that the idea of the Anthropocene is now in vogue, or perhaps its currency is no longer ascendant, or its definition and usefulness as a term is in question.[3] Instead, focus on the intention behind the term. A consortium of global change scientists use the term, and indeed invented it, to claim that human impact on the biosphere qualifies as a significant geologic force in its own right. Hence we enter a new geologic era, the Anthropocene, to denote the necessity of understanding that impact. In effect, the term becomes a code word for calling attention to exceedingly rapid global environmental change. If you study the Anthropocene literature, it reveals several interesting controversies. Scientists debate when it actually started. Was it with the steam engine? Was it due to human agriculture? Environmental philosophers ponder the conceptual ramifications. Is it yet another example of human hubris—another way we declare our superiority? Or is it a legitimate call to pay closer attention to the responsibilities of stewardship? Ghosh, the great novelist, in his insightful book about climate change, *The Great Derangement*, suggests that the Anthropocene is the biosphere's way of reasserting its control, as if it were a sentient force striking back at human foolishness and neglect.

Before the Anthropocene concept sprouted (in 2000), scholars were accumulating copious evidence about the ramifications of that geologic impact. In the European intellectual tradition, Alexander von Humboldt launched this inquiry (among many other subjects) with the 1845 publication of *Kosmos*, followed by George Perkins Marsh's *Man and Nature* in 1864.[4] While Charles Darwin was pondering the origin of species (1859), Humboldt and Marsh were wondering how species' habitats were being systematically denuded. Could we say, then, that the Anthropocene begins when the Western intellectual tradition first discovers human environmental impact?

Many other landmark works follow. Eduard Suess coined the term "biosphere" in 1911, roughly a century before the Anthropocene concept entered the scene. Vladmir Vernardsky's great work, *The Biosphere* (1926), established the significance of biogeochemical cycles, thus asserting that life itself in its multitudinous forms is a geologic force. Let us remember, too, that microorganisms have always been the primary life force driving rapid environmental change. By midcentury, promoted by the rapidly expanding human footprint, comprehensive earth system assessments emerged, now requiring teams of multiple, interdisciplinary researchers, culminating with great anthologies such as *The Earth as Transformed by Human Action* (1990), and the International Geosphere-Biosphere Programme's magnum opus, *Global Change and the Earth System*.[5] One hundred and fifty years later, these massive research programs essentially proved what Humboldt observed and intuited: humanity must come to grips with rapid global environmental change. In the last several decades, the precision of scientific research has dramatically improved monitoring capacities, data collection, and algorithmic scenarios. This exceptional data array is available to anyone who wishes to pay close attention.

Yet for many reasons, it's exceedingly difficult for people to grasp the enormity of this challenge. I cover this topic at great length in *Bringing the Biosphere Home* (2001) and will return to some of those themes in this book. Despite the best efforts of global change scientists, environmental policy advocates, environmental educators, and a handful of visionary political figures who have made this a priority for leadership, insufficient attention is paid to human impact on the global environment.

There are positive trends that we should note and support, though. There is now a global green business network—largely corporate, but also including small businesses—that emphasizes recycling, supply chains, reducing toxicities, promoting energy efficiency, and carbon emissions reduction.[6] Surely this is profit oriented and branding conscious, as Naomi Klein (*This Changes Everything*) deftly points out, but it is nevertheless significant and impactful. Who could have predicted there would be a global network of sustainability managers working in multinational corporate offices? In the public sector, there is a global cities network of sustainability managers that coordinate urban sustainability initiatives, share best practices, and strive to develop sustainable efficiencies, not only for investment purposes, but to support quality-of-life indicators.[7] This is also a relatively new development, not much more than a decade old. Below the nation-state level, in both cities and corporations, and certainly at the grassroots level, there is an enhanced urgency to reduce human impact.

Thankfully scientific research continues apace as well, despite the anti-science rhetoric of reactionaries and deniers. Notwithstanding the most rational efforts to prove the significance of human impact, the challenges of rapid environmental change remain hard to grasp, rarely covered by the dominant channels of the news cycle, emerging only when a serious storm or catastrophe hits, and then quickly forgotten when the news cycle loses interest. Those impacted by the disaster, however—the dispossessed Ghosh refers to—live with the consequences, and must find ways to internalize and respond to what happened. They must rebuild their lives or move elsewhere while living on the front lines of global environmental change.

An Intricate Convergence

The exiles' hope: to be forgotten is to become invisible, but to become invisible is not to be forgot.
—Kanishk Tharoor, *Swimmer among the Stars*

According to the 2019 Global Report on Internal Displacement, there were 28 million new displacements in 2018. The report refers to displacements as people who are forced to leave their homes because of environmental disasters, political violence, or other factors. An estimated 41.3 million people were living in internal displacement—the highest figure ever recorded.

The 2019 report suggests that this all-time high is the result of sudden-onset hazard events. Heightened vulnerability and exposure to sudden-onset hazards, particularly storms, resulted in 17.2 million displacements in 144 countries and territories. The number of people displaced by slow-onset disasters worldwide remains unknown as only drought-related displacement is captured in some countries, and only partially. In addition, "urban conflict triggered large waves of displacement and has created obstacles to durable solutions."[8]

In 2017, extreme weather events caused catastrophic floods in Houston, Texas, and levied enormous infrastructure damage on Puerto Rico. In 2018, global wildfires were a primary cause of displacement. Every nation is vulnerable.

Beyond Borders, a 2017 study compiled by the Environmental Justice Foundation, cites senior US military and security experts who claim that the number of anticipated climate refugees in the next decade will "dwarf those that have fled the Syrian conflict, bringing huge challenges to Europe."[9]

These displacements spark a sequence of political, economic, and social challenges. In Europe, the refugee crisis deepens controversies regarding border security and humanitarian concerns, spawning reactionary responses, often with distressing racial overtones. In the United States, similar fears, wrapped in paranoia about terrorism and immigration, prompt hysteria for a border wall, catalyzing racist propaganda and mean-spirited policies. The polarizations around this issue, fueled by insider/outsider dichotomies and manipulated by demagogues, frequently prevent serious policy discussions. These are important and difficult public policy questions, but they are easily subverted by centuries-old racial fears and ignorance.

As these reports indicate, there are many reasons for internal and external displacements—from climate catastrophe to political violence. There are subtler forms as well, such as dislocations that uproot citizens and communities. This might include the detritus of gentrification, abandonment of a rural community, need to leave home because there are so few opportunities, or sheer desire to move elsewhere at the prospect of a much better life.[10]

On a global scale, all these trends and patterns are meaningful. But the most serious projections of all are the chilling scenarios unleashed by climate change, with the prospect of entire regional habitats rendered unlivable, either because of a specific weather event or the cumulative changes over decades, such as desertification, flooding, or glacial melt—phenomena that creep into a community and then reach a tipping point.

These displacements, considered most broadly, weave an intricate convergence, demonstrating the interconnectedness, pertinence, and ubiquity of the tides of change. Climate change, a planetary-scale pattern, results from unmitigated natural resource extraction contributing to rising seas and catastrophic weather events. Public attitudes regarding refugees, migrants, and the dispossessed rekindle historical-scale dilemmas about insiders and outsiders, immigration, and racism. Discussions about resilience, recovery, and revitalization raise community-scale questions of economic and social justice: Who will open their communities, share their assets and strengths, and provide succor and welcome to the dispossessed? How these decisions get made is ultimately a question of both conscience and agency. At the individual scale, people make judgments that contribute to public policy. Will the approach to these decisions include spirited public debate and meaningful conversations, or will they be relegated to homeland security and border guards? All these challenges are easily manipulated through preconceptions, stereotypes, and falsifications. There are many ways that we trick ourselves, such as through confirmation bias, anecdotal information, cherry-picked facts, and closed-mindedness.

This example, the staggering prospect of climate dislocations, is enormously complex, but it does and will impact every planetary citizen. No matter how you enter the discussion, as a climate justice advocate or immigration rights proponent, if you think deeply about your concerns, the interconnectedness of these themes will emerge. Open any of these doors, and they quickly lead to the inevitable pathway of human history in the biosphere—requiring more than just an awareness of environmental sustainability, but also an understanding of the trials and tribulations of tribes, hierarchies, and citizenship.

Always the Biosphere: Environment

If you review any of the comprehensive, scientific global environmental change assessments published in the twenty-first century, you come away deeply troubled at the prospects. It's not just the threats to humanity, as severe as they may be, but also the short- and long-term ecological and evolutionary consequences of human impact. Remember, that impact is troubling enough to suggest a new geologic era—the Anthropocene.

These impacts are covered at great length in many outstanding studies and dozens of popular books. You can follow the latest scientific research in contemporary journals and on websites. Still, it's helpful to briefly review the most significant trends, serving to reinforce the necessity of ubiquitous reminders as well as reiterate how these trends impact all aspects of the human condition.

We are in the early phase of a planetary emergency, triggered by this human impact. Ultimately, human survival hinges on our ability to recognize, internalize, and then adapt to this emergency. There are four interconnected, biosphere-scale challenges—the sixth mega-extinction, habitat fragmentation, rapidly changing oceanic and atmospheric circulations, and altered biogeochemical cycles. Let's briefly review each challenge.

Evolutionary ecologists are amassing sufficient data to determine that the biosphere experienced five previous mass extinctions, and we are now entering the sixth one. A mass extinction is loosely defined as a geologic-scale event in which over two-thirds of the earth's species perish. Previous extinctions were the end of the Ordovician (444 million years ago), late Devonian (375 million years ago), end of the Permian (251 million years ago), end of the Triassic (200 million years ago), and end of the Cretaceous (66 million years ago). Scientific speculations as to their causes include severe planetary cooling and heating, rapidly changing atmospheric chemistry, powerful microbial metabolisms, and catastrophic meteor strikes. The full explanations for these extinctions are still being explored and discovered. Many ecologists and conservation biologists warn that a sixth mass extinction is now underway, or as reported in a recent study from the National Academy of Science, "Dwindling population sizes and range shrinkages amount to a massive anthropogenic erosion of biodiversity and of the ecosystem services essential to civilization. This 'biological annihilation' underlines the seriousness for humanity of Earth's ongoing sixth mass extinction event."[11]

Unique to the sixth mass extinction is the hypothesis that human agency is the direct cause, raising complicated as well as challenging ethical and moral questions, all of which should inform political and economic decisions. A recent study suggests there are 8.7 million species on earth, but that number is contested by microbial ecologists, who maintain that the number is far greater. E. O. Wilson reminds us that we have greater precision in identifying the number of stars in our galaxy than we do about the number of species on earth. He reminds us, too, that each species represents an encyclopedia of evolutionary knowledge, and each species lost is akin to burning volumes in a library.[12] Needless to say, the world has thousands more lawyers and bankers than it has soil scientists, and as long as that's the case, our ability to estimate these numbers will remain iffy. In the long run, a resilient biosphere will evolve accordingly, but unless you plan to live several million years, this fact won't help humanity or the legions of species that will go down with us.

One of the main causes of species extinctions is habitat fragmentation, mainly caused by the rapid spread of global urbanization and agriculture, carving the landscape into vast areas of human settlement and domestication. Of course, some fragmentation is inevitable with any human settlement, but we don't typically think about what happens to the great variety of species that is displaced when, let's say, we build a parking lot. Or as stated more scientifically in a landmark study, "Habitat destruction typically leads to fragmentation, the division of habitat into smaller and more isolated fragments separated by a matrix of human-transformed land cover. The loss of area, increase in isolation, and greater exposure to human land uses along fragment edges initiate long-term changes to the structure and function of the remaining fragments."[13]

There is much ado about climate change, and public awareness sparks after every hurricane, drought, or catastrophic wildfire. I prefer to conceptualize climate change in a broader context: rapidly changing atmospheric and oceanic circulations. The earth is an intricate (and beautiful) circulatory system, with waves, patterns, and flows of energy, transported around the globe on water, land, and in the air. For a spectacular and inspiring visualization of this process, especially in the world's oceans, see the Perpetual Ocean videos at NASA's scientific visualization studio.[14] These oceanic and atmospheric flows and circulations are influenced by the much-slower but perennial movements of continental plates, and the injection of material from the earth's interior onto the surface, in deep sea vents, volcanoes, and through other permeable surfaces. There's a lot going on, and it is remarkable to behold. This is the context through which atmospheric carbon alters the biosphere, impacting multiple circulatory systems. We know that the increased amount of thermal energy in the atmosphere and oceans has dramatic biological, ecological, and geomorphological consequences—from increased photosynthesis rates to methane release to changing phenological patterns to rising sea level.

These energy flows and circulations are embedded in a network of cycles. The biogeochemical cycles—nitrogen, carbon, water, and phosphorus, and many lesser-known others—are impacted by increased thermal energy and the material outputs of human activities, notably agriculture and industrialization. Consider the extraordinary finding that various species of fish in the Great Lakes have high levels of antidepressants in their bodies. Or consider the gyres of plastic bottles, spinning a vortex in the Pacific Ocean, or toxic algae blooms in the Gulf of Mexico.

The carbon cycle is the most familiar given that measuring atmospheric carbon levels is a marker of climate trends, and mitigating those levels is the controversy at the heart of public policy solutions. These cycles, taken together, contribute to individual and global metabolisms, chemical balances, and ecological physiologies. Worries about ocean acidification are yet another way that altered biogeochemical cycles impact biodiversity and ecosystem health.

One of the great accomplishments of twenty-first-century global change science is the rapidly advancing understanding of these interconnected trends. Still, even the most engaged citizen has a hard time understanding the depth of these intricate interconnections. Nevertheless, we won't come to grips with any of these issues until we pay closer attention to them. That's at the core of our educational challenge and crucial to the future of environmental learning. Much of this book will explore educational approaches for doing so. But before we begin that process (in chapters 4 through 7), I'll resume the discussion of how these biosphere-level changes exacerbate the tides of change, and why such awareness is the key to understanding why environmental learning matters. I'll start with tribes and territories.

Tribes and Territories: Diversity

Cooperative altruism of the style we can find in our species has paid handsome dividends in our past—dividends that arise from assembling a powerful and cohesive social vehicle made up of the individuals committed to cooperating with each other. It is this that makes human culture the survival vehicle that it is, and we have evolved an entire psychology around it, from our acts of kindness and self-sacrifice to our xenophobia, parochialism, and predilection to war.
—Mark Pagel, *Wired for Culture*

I haven't lived in New York City since 1971, and I don't get to visit nearly as often as I would like. But if I meet a New Yorker, I feel an affinity. It's even better if I meet someone who is Jewish and was born in Brooklyn. Even better than that is if they root for my favorite sports teams: the New York Mets and New York Knicks. And if they share a similar political persuasion, an interest in environmental issues, or other common interests, I'm thrilled. My son grew up in New Hampshire, and I fervently hoped that he wouldn't become a Boston Red Sox or Boston Celtics fan (he didn't!). Their fans and uniforms arouse distaste. If I can reveal more of my superficial, petty, tribal side, I am more comfortable with New

Yorkers than Bostonians, and find more of an affinity with New York Jews than Boston Jews.

I have dozens of other "soft" tribal affiliations. I'm proud to live in the Monadnock region of southwestern New Hampshire and strongly identify with the place. My son once said that people who live in Vermont are afraid to live in New Hampshire. Our state motto is "live free or die." And yet I strongly identify with northern New England, including Vermont. When I lived in Seattle, I told people that I was from New Hampshire, and before that I was from Long Island and Brooklyn. My grandparents came from shtetls in Russia. Now that I'm back in New Hampshire, I feel an affinity for Seattle.

When I was twenty years old (1970), I led a group of teenagers on a bicycle trip through Europe. It was my first visit to Germany. I assumed that anyone forty years or older lived through the Nazi regime. Just hearing the language gave me shivers. I no longer feel that way, but I did then.

These impressions, affinities, and antipathies may seem silly. If they do, I suggest you organize your tribal affiliations and see where they stretch. To which ones are you most attached? Look at the dark side too, and think about those groups for which you feel animosity. Perhaps, like me, you can generate a spectrum of emotions, some linked to solidarity, and others linked to fear.

Entire elections and the fate of nations may ride on these emotions. The US election of 2016 was about building border walls, excluding persons of the Islamic faith, and preying on stereotypes of race, ethnicity, and gender. The refugee crisis in Europe and elsewhere prompts some people to extend compassion, and others to promote xenophobia and racial purity. A thin line separates the rabid fans at sporting events from the more sinister manifestations of tribal impulses.

In his outstanding book *Behave: The Biology of Humans at Our Best and Worst*, Robert Sapolsky devotes an entire chapter to "Us versus Them" dynamics.

Neurobehavioral research suggests that "the strength of Us/Them-ing is shown by: (a) the speed and minimal sensory stimuli required for the brain to process group differences; (b) the unconscious automaticity of such processes; (c) its presence in other primates; and (d) the tendency to group according to arbitrary difference, and to then imbue these markers with power."[15]

Sapolsky develops a fascinating taxonomy of us versus them dynamics, describing how we all belong to multiple categories of us, the various ways we generate those categories, and whether there are reflective processes that help us mitigate those distinctions.

The evolutionary psychology of us versus them dynamics is one foundation of tribal affiliations, territorial markers, ethnic differentiations, and cultural distinctions. Superimpose ecological restraints, hierarchies (see the next section), gossip, and other characteristics of the human condition, and the challenge of tribes and territories becomes clear. If you personalize it, as I suggested above, you can see how easy it is to develop us versus them distinctions. Multiply it by history and culture, and see where it takes you.

We should consider the extent to which scarcity conditions, catalyzed by rapid environmental change, exacerbate or mitigate us versus them dynamics. Sapolsky, Pagel, and Joshua Greene, who all cover the relationship between tribal affiliation and moral behavior, assert that depending on factors such as affluence, equity, emotional intelligence, and impulse control, humans are capable of a wide range of responses—from compassion to exploitation to expulsion.[16] Most important, they also suggest that education and reflective awareness may contribute to constructive approaches. I will return to those hopeful prospects later and frequently, specifically in chapters 6 and 7.

Earlier in this chapter, I presented an argument that weather-related events (a subset of rapid environmental change) create dislocations and displacements, or scarcity conditions. Inevitably, our responses to these situations are framed by us versus them dynamics. Hence environmental learning must address the relationship between tribal affiliations and global environmental change. Human history is replete with tribes and territories, and as planetary settlement becomes comprehensive and interconnected, boundaries blur and territories superimpose, while diverse cultures increasingly meet face-to-face. Walls will not prevent this.

Sapolsky observes that "from massive, breathtaking barbarity to countless pinpricks of micro aggression, Us versus Them has produced oceans of pain."[17] We will never be cured of us versus them. It's too deeply ingrained. But it can be mitigated. That is a task for environmental education because as we better understand these impulses, we enhance our capacity to adapt to rapid environmental change.

The Stress of Inequality: Equity

> Put simply, cultures with more income inequality have less social capital. Trust requires reciprocity, and reciprocity requires equality, whereas hierarchy is about

domination and asymmetry. Moreover, a culture highly unequal in material resources is almost always also unequal in the ability to pull the strings of power, to have efficacy, to be visible.
—Robert Sapolsky, *Behave*

As you cruise through your day, you engage with many different groups, organizations, and institutions. Depending on your relative stature and experience in any of those groups, regardless of their governance structures or informal social arrangements, you will have a relative ranking in a hierarchy. You can be a minimum wage cook and the captain of a softball team. As Sapolsky points out, it takes only forty milliseconds for the brain "to reliably distinguish between a dominant face with direct gaze) and a subordinate one (with averted gaze and lowered eyebrows)."[18] Rank and hierarchy often lead to inequality. These relationships appear throughout human history, taking various forms depending on ecological and cultural circumstances.

Historical evidence indicates, as Kent Flannery and Joyce Marcus make clear in their exhaustive study, *The Creation of Inequality*, that the institutionalization of inequality accelerates with the development of agriculture, origins of surplus, and requirement to organize that surplus. Sapolsky claims that "inequality expands enormously when cultures invent inheritance within families." Most significant, "once invented, inequality becomes pervasive."[19] Of course, inequality further enhances the stratification of rank and hierarchy leading to structured dominance loops in states, bureaucracies, and corporations. Inequality manifests in multiple forms, especially in the distribution of and access to ecosystem services, availability of social capital, and opportunity to advance in cultural ranking systems.

Thomas Piketty's landmark study, *Capital in the Twenty-First Century*, presents copious evidence that over the last two hundred years, increasing concentrations of wealth have resulted in great disparities between rich and poor. Piketty expects this trajectory will continue in the twenty-first century unless redistribution policies address this egregious pattern. Indeed, the challenge of wealth inequality is a prominent controversy, and political polarization emerges from the vastly different solutions for addressing it—stated most simply as tax cuts versus entitlements, or supporting unfettered business growth versus state-organized redistribution.

The equity challenge transcends wealth accumulation per se. And it's crucial for how we think about environmental issues. The US Environmental Protection Agency developed an environmental justice mapping tool, called EJSCREEN, available for public use, clearly demonstrating that toxic waste sites are located closest to low-income regions, especially those inhabited by communities of color, and parks and green spaces are located closer to affluent neighborhoods.[20] For an outstanding illustration of these dynamics from an international, cross-cultural, and historical perspective, see the online Environmental Justice Atlas.[21] These tools, combined with hundreds of environmental justice studies, show the clear link between wealth inequity and access to ecosystem services. As the displacement studies described earlier demonstrate, these challenges will become more severe as weather-related events and other manifestations of rapid environmental change most prominently impact poor or disenfranchised communities.

Inequality contributes to stress in three powerful ways. First, there is the sheer psychological dynamic of status deprivation, low rank in hierarchies, and perceived unfairness that accompanies those situations. Second, stress arises from the tangible material disadvantages of limited access to goods and their distribution. Third, the pervasive distrust caused by inequality impedes the very social capital formation that provides mutual aid and encouragement. These are all measures of both community and individual well-being.

If you investigate any environmental issue, you will inevitably confront the challenge of inequality. And if you are mainly concerned with inequality, you will inevitably understand how and why some communities have unequal access to ecosystem services. Layer those concerns with the us versus them challenges portrayed above, and you get a more complete picture of why environmental learning must involve all these factors. Ultimately the solution to these issues depends on governance, decision making, and social mobilization, and so that is where we now turn.

Participation and Democracy: Inclusion

Democracy is at once the most serious obstacle those who would address climate change must overcome, and their indispensable vehicle for achieving success. Democracy corrupted turns mass opinion toward narcissism and self-interest. Democracy actualized and legitimate offers a politics of hope against corrupted democracy's politics of fear.
—Benjamin R. Barber, *Cool Cities*

Several decades ago (1986), Barber wrote a magnificent book: *Strong Democracy*. It advocates a paradigmatic shift in how we think about governance. In Barber's view, thin democracy emphasizes representation. A few people (elected representatives) make all the decisions all the time. Strong democracy stresses participation. All citizens make some of the decisions some of the time. Barber has an outstanding concluding chapter in which he presents a catalog of options for implementing strong democracy.

I first encountered Barber's work in my thirties. I was developing a variety of environmental studies courses. I knew that for the field to be effective, it must challenge students and citizens alike to deepen their understanding of political agency, and empower them to take action. In an earlier book, *Ecological Identity* (1995), I developed a suite of learning activities to help students better understand their political values, particularly in relationship to property, community, and decision making. A strong ecological identity (identification with species, habitats, and the biosphere) should be the values foundation that inspires political action. Your political temperament (linked to personality traits) helps you determine the forms of action that are most suitable, and how you can be most effective in public, organizational, and institutional settings. If you are an environmentalist, then you must also be a citizen. How will you participate in public life?

Thirty years later, in *Cool Cities*, Barber examines the politics of climate change, weaving his concern about its potentially catastrophic ecological consequences with his awareness of social justice, culminating with pleas for public action.

Democracy is crucial because climate change is also about justice: how to distribute the costs of decarbonization and transition to renewable energy equally as well as fairly among rich and poor, developed and developing, large and small, north and south. We cannot elude the subject of justice because the costs and benefits of addressing climate change are inevitably skewed around wealth and power.[22]

Barber, like many of us, is dismayed at "the seductive demagoguery of populist pretenders" who preach a "politics of cynicism" while transforming the aspiration for inclusion into authoritarian impulses. He understands the frustration with global elites, political bureaucracies, and dysfunctional ideological stalemates perpetrated by oligarchs whose wealth provides disproportionate access to the halls of national and global power. Barber explains how easily democratic aspirations are undermined by "money, media, and manipulation," and how a thin line separates the desire for participation from the political drug of addictive nationalism.[23] Gary Snyder, the great poet and essayist, describes nationalism as "the grinning ghost of the lost community."[24] How do we overcome the false promises, delusional divisions, artificial racial constructions, and desperate clinging to an idealized past that never really existed?

In both *Cool Cities* and *If Mayors Ruled the World*, Barber offers political solutions and opportunities based on the capacity of cities to address real-world problems in a practical, nonideological, inclusive way. He presents evidence suggesting how this is happening in towns and cities around the world, and how these principalities are forming constructive networks of empowered action. I will elaborate on these trends in chapter 7, "Cosmopolitan Bioregionalism," while discussing the relationship between the local and global.

The tides of change unleash waves of psychological and political uncertainty. Displacement and dislocation upend the habits of routine and normalcy. In times of uncertainty, the dangers of authoritarianism rise to the surface, and the prospects for democracy are more easily threatened. As rapid global environmental change impacts more communities, these tensions will become more prevalent. Do we choose authoritarian solutions or participatory decision making? Do we choose deliberation and inclusion, or unilateralism and exclusion? Do we find measures of compromise and common ground, or do we resort to coercion and violence?

Openness and Vigilance

It's challenging and essential to contemplate how the tides of change form connected loops of impact. The intellectual exploration of their connectedness is intricate, interesting, and instructive. But how do you respond when you experience the displacements and dislocations that may follow the tides of change? What happens when it's personal and becomes your reality? You abandon your home when the rising tide flushes it to sea. You seek a new life in a foreign country only to be turned away at the portal of entry. You are denied opportunities because of your cultural background or the color of your skin. Your lifelong place of employment is closing its doors and moving to another country where the cost of labor is much cheaper. How can you possibly maintain clarity about what's happened to you, when all that you've known shatters and dissipates?

And then, what are your feelings and opinions when you are an outsider looking in—a spectator who observes such dislocations by watching the evening news, following a Facebook feed, or chomping the bits and bytes of Twitter? Unfortunately, public awareness of displacement or dislocation is often indirect, thirdhand, or subject to the whims and frenzies of pundits and media personalities. You wonder how to generate empathy for the tribulations of others when you have problems enough of your own, or whether there is room enough in your town, community, or nation for people arriving on your doorstep.

The US presidential election of 2016 was presented as a referendum on these issues. Consider the epithets and chants: "build the wall," "lock her up," and "make America great again." But on the day after that election, despite all the concerns and worries, the change maker projects I described previously were still doing their work. The community-based efforts were just as challenging and rewarding as they were the day before the election. The obstacles appeared greater, especially with the decline in federal funding initiatives. And communities of color, immigrant families, and all the disenfranchised became more vulnerable and vigilant. Yet people I spoke with (to this day) emphasize a new resolve, compelling solidarity, and even deeper love for their work in community. They express an openness to work even more closely together, strengthen their ties, and reach out in new as well as more innovative ways.

Openness and vigilance. That's what I felt after the US 2016 election. Is it possible that the anticipatory threats we fear might present new possibilities for wisdom and action? Or that we'll reach new levels of compassion and empathy? As we deepen our awareness of the tides of change, we have no choice but to express renewed commitment and vigor, a belief in people, and a desire to be more open and understanding. We should never underestimate the devastating possibilities, necessity of vigilance in the face of ignorance and prejudice, tangible threats to communities of color and the ecosystem, and prospects of deepening divides amid the clouds of uncertainty.

Activists and educators face urgent challenges. We start in our classrooms and communities, in our daily conversations and ambient encounters. Nothing matters more than engaging in meaningful conversations about things that matter. We don't have to agree. But we can listen and understand. This isn't a naive dream. I'm well aware that there are some who won't listen and are happily arrogant in their views. The dangers are as real as the opportunities. That's why we must stay open and vigilant. The work continues. It always has.

All four of my Jewish grandparents left Russia in fear of pogroms. They made better lives in New York City. They were open to new possibilities just as they stayed forever vigilant. One hundred and ten years later, our prospects are simultaneously enhanced and diminished. There are always more lessons to be learned. And as ever, our learning begins anew. That's what makes for successful projects, and that's how we restore our communities.

The range of human responses to the tides of change is as varied as the spectrum of human behavior—from indifference to empathy, from belligerence to compassion, from denial to engagement. It is way beyond the scope of this project to explain, elucidate, or elaborate on why people respond the way they do. Rather, I wish to convey three points here, and will pursue them more closely from various perspectives throughout the book. First, to deeply understand the tides of change, we require both an intellectual perspective and the self-awareness to reflect on our emotional responses. Second, the tides of change are not unique to the Anthropocene. The consequences of environmental dislocations are a fundamental aspect of human evolutionary ecology. The challenges of sustainability, inequality, tribalism, and participatory decision making are intrinsic to all of human history. Third and most important, a deeper understanding of this legacy requires an educational response—one that emphasizes the interconnectedness of these challenges. Environmental learning is only pertinent when it demonstrates the intricate convergence manifested by the tides of change.

It's the Human Condition, Isn't It?

Very deep is the well of the past. Shall we not call it bottomless?
—Thomas Mann, *Joseph and His Brothers*

The tides of change may be flooding the Anthropocene, but they are intrinsic to all of human history. The challenge of sustainable resource use is a pervasive ecological dynamic. What resources does a tribe, nation, or culture require, and what strategies does it choose for procuring them? From the dawn of prehistory, human cultures have wandered the globe. They develop strikingly unique cultural forms, and one never knows what happens when they encounter each other. Will it be peaceful coexistence or endless aggression, or perhaps something in between? To manage the ecosystem, distribute wealth, and assess external encounters, humans develop hierarchies and decision-making systems. Who gets to participate, and who attains power?

There are countless cultural and organizational forms depending on ecological context, biological proclivities, and the transmission of social learning. The cumulative human experience is staggeringly complex, yielding languages, art forms, technologies, and a wide variety of ecological and cultural practices. In the last decade, with the emergence of big history, evolutionary psychology, ecological theory, and wonderful new forms of interdisciplinary scholarship, there's been a proliferation of data, interpretation, and theory that addresses the human experience in an evolutionary context.

As Sapolsky makes clear, it's not useful to attach causation to these phenomena. Instead, we're best served by exploring the complexity of diverse variables influencing each other. Beware, then, of ideologues who extrapolate their findings to create elaborate rationalizations or causal sequences that promote a specific point of view. Remember, too, that human behavior stretches into the dawn of time, and our best and worst impulses precede modern history. Stay open to cultural forms and ecological approaches that promote awareness, mindfulness, love, and compassion. Be cognizant that the best intentions are easily misconstrued and our biological origins can easily descend into raw, unexamined, dangerous behaviors.

What does all this have to do with environmental learning? I'll start with sheer practicality. If you believe strongly that climate change requires urgent action, you have to convince others of the necessity of action and the specific requirements of policy approaches. At every stage along the way, you will encounter people who at worst deny the issue entirely and at best disagree with you about political strategies. Discussions easily degenerate into us versus them stereotypes, miscommunications, and distrust. What are your alternatives? You can try to find common ground with those whom you disagree. Sometimes that's possible. Sometimes it isn't. You can try to mobilize people who think similarly to you, and do your best to consolidate as well as project political and moral power. None of this is possible unless you cultivate an understanding of how us versus them dynamics are rooted in the human condition. A rational argument for mitigating carbon emissions might not work, despite your best ability to communicate the consensus view of the most statured climate scientists. With hard work, good listening, and respect for the complexities of human behavior (including your own), you might be able to transcend us versus them polarizations, and find some way to move forward.

Let's shift this example into a more emotional realm. How does a nation respond to climate refugees? If it responds with the worst us versus them polarizations, it will create walls and boundaries, develop stereotypes of the victims, rationalize their plight by saying it wasn't our fault, and respond with the cruelest possibility of all: indifference. Again, I rely on Sapolsky's advice.

Distrust essentialism. Keep in mind that what seems like rationality is often just rationalization, playing catch up with subterranean forces that we never suspect. Focus on the larger, shared goals. Practice perspective taking. Individuate, individuate, individuate.[25]

What's the best way to individuate? Follow the narrative, life stories, and human circumstances of the refugees' plight. How do you enhance perspective taking? Ask what you would do if you were in their shoes. Think of the similarities between their circumstances and your own, and remember how a simple twist of fate could determine the timing and sequence of dislocation.

Please recall the fog man story in the first chapter. A colleague at New York University asked her environmental education class to read an earlier version of that memoir. The students posted their responses on my website and asked for my comments. Amazingly, the story of romping through the fog captured the attention of all the students, evoking memories of similar experiences—smog in China and India, pesticides sprayed on the family lawn, and toxic waste on a playground in southern Mississippi. It's the details of the story that allows for perspective taking, permits us to individuate, and evokes insight—promoting questions, ideas, research possibilities, and approaches to collective action. The stories bring us closer to the lifeblood of human experience.

To better understand the tides of change, we must emulate the best examples of human behavior—the cultivation of empathy, altruism, and compassion. No one said it would be easy. But there is no greater task for education. It requires discussions of virtue and intent, potential and possibility, power and privilege, insiders and outsiders, and the individual and the collective, all formulated within the context of human evolution in the biosphere. These discussions, as challenging as they may be, are a foundation for evocative, resilient, and enduring environmental learning.

The Anthropocene tides of change occur in an extraordinary cultural environment, catalyzed by the rapid transmission of information, conceptual intellectual breakthroughs, a profound shift in how people learn, the immediacy of their daily experience, and how they come to know the world. The next chapter examines the opportunities and perils that lie so close at hand, and offers some suggestions for how they can best inform environmental learning.

4

Is the Anthropocene Blowing Your Mind?

Screens, Windows, and Frames

Nadia and Saeed were, back then, always in possession of their phones. In their phones were antennas, and these antennas sniffed out an invisible world, as if by magic, a world that was all around them, and also nowhere, transporting them to places distant and near, and to places that had never been and never would be.
—Mohsin Hamid, *Exit West*

If it is true that a picture is worth a thousand words then what is the power of the billions of images that now permeate every corner of the globe? What is the potency of the dreams and desires they generate? Of the restlessness they breed?
—Amitav Ghosh, *Gun Island*

As I write this passage, I'm staring intently at a MacBook Pro, my entire world seemingly channeled onto a screen about two feet from my head. The deeper my concentration, the more intense is my gaze. When I get distracted, I travel to other places on the screen. Perhaps I'll check my email. Or maybe I have an idea regarding something I'm writing about, a "frame" of reference, and I choose to surf the web to learn more. I'm well aware that although I seem to be guiding my journey, whatever search engine I'm using is subtly steering me. I also know that my web journey may be tracked by anyone who chooses to look into my activities. Yet I still feel that I have autonomy and am in control of the internet places I visit. I know, too, that there is an addictive quality to this journey. I must exercise self-control and discipline to avoid the endless distractions that may feel serendipitous, but are nothing more than dopamine pleasures.

I turn my attention away from the screen. I look around my study. It's a small room filled with books. There are approximately one thousand books, separated into sixty partitioned shelves, with each shelf a frame that organizes the books into subject areas pertinent to my interests. Every volume is a new journey, a world unto itself, and a possibility for further exploration. Each volume also triggers memories of what I was thinking about when I acquired the book and why I thought it might bring meaning to my inquiries. Browsing these shelves is a source of inspiration as well as a trip down memory lane. I'm constantly reminded of the extraordinary ideas these books contain and great minds that thought them. I consider the time and attention I spend curating this collection. It allows me to wander through ideas within a conducive framework for how I think and what interests me. My attention shifts between the computer screen and the books on the shelves in my study.

On my desk, there's an old shortwave radio. It picks up every conceivable transmission from countries around the world. Mostly I use it to listen to National Public Radio or perhaps a ball game. It seems so old fashioned. Yet at one time, in the days before the internet, my radio was a portal to the world, with each channel a distinct acoustic/information space, in many cases providing unique perspectives on news and culture. The radio is also a beacon of memory. I can tell you where I was the first time I heard the Byrds rendition of "Turn! Turn! Turn!" I vividly remember (as a five-year-old) listening to the fabulous Vin Skully announcing Brooklyn Dodger games or (as a twelve-year-old) the first-ever New York Mets game. Just by looking at the radio I conjure dozens of memories, each serving as access points to moments of learning.

The radio channeled sound into discrete packages of news, sports, and entertainment, framing acoustic space, allowing the listener to travel through prepared narratives. Television combined visual and auditory stimulation, providing even more prepared content. In the beginning of each medium there were just a few choices, often managed by the state or else corporations that could afford the technological investment. As radios and then televisions became less expensive and more accessible, the choices proliferated, allowing for more independent voices and the ensuing competitive demand for public attention. They paved the way for the internet, combining sound, video, and instant two-way communication, transforming attention spans, cognitive perception, and the speed of information transmission. Compare television programs or movies from even a decade ago to current content, and notice how the frame rates have changed. Consider your impatience when a website isn't loading fast enough. We now expect to observe multiple screens (frames) simultaneously. How much more speed are you accustomed to? And how much can you take in?

Next I look out the south-facing windows that bring light into the room and reveal the great outdoors. I can only see what the window frame allows: a discrete view into the New Hampshire woods. Today, a late March day, the woods are still snow covered. It's a cloudy and cool forty-six degrees at noon. I take off my glasses (yet another frame) and my focus expands. The window is closed so I can't hear or smell much beyond the confines of my study.

As I proceed through this exercise, and shift my gaze from the computer screen to the book shelves to the window and then back again, I rapidly shift frames. Hooray! I am multitasking. I am talking to you the reader, my invisible friend, my anonymous audience, accessed through these pixels on a screen eventually to become a book (which you are now reading) while simultaneously scanning my shelves and looking out the window. I can barely take it all in. I can do all this while sitting in my chair.

I forgot to tell you about the photographs and paintings that fill the spaces between the shelves. There are pictures of my family, artworks from family members, and a few small scattered paintings. There are several dozen rock and mineral specimens and fossils, beautiful crystals from various mines around the world, treasures from the earth's underground, each telling an extraordinary geologic story about minerals and life, now sitting on these curated cabinets.

I look toward the right through the door of the study into our bedroom, another container of memories, curiosities, and experiences. The open doorway is a passage from one state of mind to another. I venture downstairs and go outside so I can get more deeply immersed in the outdoors beckoning from the window. I step outside and feel instantly liberated. The snow and gray sky are so bright and open. I can hear the drips of water from melting snow on the roof. I observe the patches of bare ground, wondering when and where the first crocus will sprout. I walk around the house, through both snow and open ground, snagged by a wayward blackberry branch. I hear a chickadee in the distance. I detect the scent of old grass along with a few other olfactory sensations emerging from the snow and ice. There's lichen and moss on the trees. The world is now alive. I'm out of the box and in the biosphere. What an exhilarating feeling! I'm using my senses to explore the great big unenclosed world.

In just a few short moments, my awareness shifts depending on the frames I use to organize my experience. It takes a special effort to summon mindfulness regarding these shifts. As a thought experiment, I encourage you to engage in a similar mind journey. Try moving between screens, windows, and the great outdoors. Along the way, notice all the frames that enclose your world—photographs, paintings, bookshelves, doorways, and storefronts—and how seamlessly you move through and between them without thinking. Notice the way these frames expand or diminish your experience. How do the shapes and sizes of the frames change what you see as well as how you think? Notice, too, how they shift the speed and pace of your observations.

Be sure to do this outdoors, whether on a city street, in a garden patch, or on a walk in the park. Look at the sky. Look at the horizon. Notice the many species of lichen on trees or boulders. Look at a puddle. Watch a bird move from place to place. Follow an ant as it moves across a patch of ground. Consider the juxtaposition of these frames.

Now what happens when you look at your cell phone while you're walking? You've reduced the entire world to a small screen. But within that screen you can enter through a sequence of portals that will continue to reframe your attention. Observe how your attention is literally funneled into a pocket-size screen that takes you into a world of contacts and virtual possibilities.

Notice, too, the difference between active and blank screens. A screen is dark until you activate it and then it comes alive because you turned on the switch. It calls you in and becomes your personalized electronic sense of place. The screen is an interface and container. It induces a spatial boundary although it appears to be expanding the places you can visit. It mediates the experience, changing its boundaries, altering the perceptual field.

It's challenging to understand how your experience may be altered when you are fully engaged in the experience itself. No matter how much you read or think about the nature of these changes, you don't necessarily know that they are also happening to you! The habits and routines of daily life sneak up on you, and before you know it your world is always framed by a screen.

Screens, windows, doorways, and frames serve as portals that take you from one experience into another, leading you into potentially vast chains and sequences of both visceral and virtual encounters. Ask yourself where the portal takes you, and if it's really where you want to go—and perhaps, who may be watching you go there.

Autonomous and Autonomic

Our relationship with our planetary home involves meaning and purpose as well as material resources. What is it like to live in the self-created circumstances of human empire? How does it work? How does it feel?
—Rosalind Williams, *The Triumph of Human Empire*

As frame rates increase and screens proliferate, our world becomes a spirited and frequently anxious competition between compulsion and intention. This isn't unique to the internet era. You can walk through an ancient marketplace in an old city and experience a similar tension. It's the intensity and ubiquity of the challenge that now matters. We all recognize the overwhelming influence of social media, tangled webs of interactive connectivity, and changing perceptual prospects of unwitting attachment to those processes. The Anthropocene represents the acceleration of two mutually conducive trends: the global extraction of natural resources and intensity of electronic connectivity. Few spaces on the globe are spared.

So I ask, Is the Anthropocene blowing your mind? This 1960s' metaphor, originally referring to psychedelics, brilliantly envisioned the future. It connotes the rapid pace of perceptual change accompanying globalization. It conjures both an enthusiastic embrace of this change, especially regarding expanded consciousness, while acknowledging the risks of taking on such a challenge.

The ecological ramifications of the Anthropocene reflect the accelerating pace of human impact on the biosphere. The psychological ramifications reflect the accelerating pace of human electronic connectivity. Our challenge is to better understand how these trends impact our psyches. My concern is that the pace of connectivity casts a spell on our attention spans, widening the gap between our psyches and the biosphere, depleting both our ecosystems and direct experience of the nonhuman. My educational agenda is to narrow that widening gap. I believe that the process of environmental learning as presented in this book is an important means for doing so.

I'm suggesting that we step back and reflect on our immersion in screens, and consider that immersion as the residing perceptual experience of the Anthropocene. Our challenge is to sustain balancing processes that provide reflective guidance for this engagement. I'll start by investigating two words—autonomous and autonomic—and how they help us understand the Anthropocene. Then I'll elaborate on what I mean by the "collective spell," and how it casts its power and leads to the seductions of ubiquitous novelty. What does it take to reclaim attention, resist compulsion, and clarify intention? And I'll consider some environmental learning tools (the balancing process) that provide educational guidance.

Consider the thin line that separates two etymologically connected words. "Autonomy" speaks to independence and freedom, and the capacity to make an informed, deliberate decision. It's a cherished value of democratic governance, human rights, and civic participation. "Autonomic" is a biological term that refers to the involuntary, unconscious activities of the central nervous system—bodily processes that occur below the level of conscious awareness. How fitting that changing just several letters in a word can make such a difference, perhaps suggesting that a fine line separates these two conditions.

The Anthropocene represents the autonomic unfolding of human impact on the biosphere, the unintended ecological consequences of expanding economic and technological capacity. Agricultural and industrial societies were adaptations to material conditions; they were attempts to provide energy and food resources that enhanced human flourishing by maximizing production. When James Watt perfected the steam engine (an invention that some describe as the birth of the Anthropocene), he was not thinking about changing the earth's carbon budget. It took two centuries of industrial expansion before the Keeling curve revealed the consequences of this invention.[1] I do not mean to imply that the rapacious extraction of natural resources is somehow innocent or corporate industrialists are unaware of the implications of their actions. Rather, I'm generalizing about long historical processes that transcend the agency of individual actors.

The Anthropocene concept is an autonomous response to this historical process, reflecting a cumulative awareness that humans must consider the wider ecological impacts of their actions and recommending a deeper understanding of intention as a means to promote ecological awareness. Environmental learning requires a mindful deliberation of resource use and impact, assuming that a more intimate awareness of natural history and ecology promotes that deliberation. The contemporary sustainability movement is founded on that premise—cultivating intention regarding human impact on the biosphere. The dilemma is that global economic exchange is so complex that it's difficult to separate the daily process of earning a living, or the relationship between enjoying consumer wealth and its inevitable ecological impact. That's why environmental learning is so hard to convey.

The internet represents the autonomic unfolding of human communication in the biosphere. The extraordinary and intricate network of global communications and connectivity accompanies expanding technological capacity. The use of this system creates profound psychological impacts. From the telephone to the television to the computer, culminating with the internet—all these innovations were designed to facilitate human communication and flourishing. The cumulative impact is a dynamic perceptual reconditioning of human sensory awareness. Although the controversies surrounding these innovations are well known (I remember the discussions in the late 1950s about television's impact on reading), the expansion of this capacity continues unabated, as an autonomic process of the Anthropocene, cumulatively reaching the deepest recesses of human communication. The perceptual impact is complex, and it's difficult to understand how the seemingly innocuous act of sending a text or email, or surfing the web, can send you down a rabbit hole and change the way you learn about the world, without ever thinking about what may be happening to you.

These convergent processes, the unforeseen consequences of daily consumer behaviors and internet activities, may be insidious and invisible, or may generate compulsive behavior—the lust for access—but they aren't irredeemable or total. Otherwise we couldn't have this conversation. The Anthropocene can also be envisioned as a time of human awakening, an era when scientific knowledge and the prospects for global awareness are unprecedented, and a period for learning opportunities and discovery. The tools are at hand. The expanding technological capacity that leads to the Anthropocene presents the potential for cultivating intention, promoting autonomy, and allowing us to contain what we've unleashed too. Before we explore that potential, let's better understand the power of the collective spell.

A Collective Spell

Procrastination is the mirror image of addiction; both are disorders of self-regulation. We are stuck in these cycles unless and until we can find the crossed signal and switch to begin again.
—Stephen Nachmanovitch, *Free Play*

On countless occasions, I've opened my computer for an assortment of tasks—to check email, visit a website, or work on a project—only to be distracted by a side trip down the rabbit hole of the internet. I may be interrupted by a text message or some other "alert." Before I know it, I've journeyed much too far from my original intention. It takes great concentration, focus, and restraint to stay true to my plan. I find, too, that as the speed of my surfing increases, I have less patience to stay put electronically. The combined internet-computer medium accelerates my brain and changes the pace of my thinking. It's a form of compulsive behavior.

I often wonder how six decades of exposure to rampant consumerism subtly prepared me for this compulsion. The great lure of consumer capitalism is the promise of unlimited bounty. Advertising presents relentless exposure to things you didn't know you needed, the means to acquire these things whether you can afford them or not, and the urgency to "act now" on your impulses. It takes great restraint to resist the temptations of your next consumer obsession. Few people are immune to this insidious cycle of temptation and reward.

The internet is an ideal milieu for supporting the relentless quest for both consumer satisfaction and social connection. It merges these impulses into a steady stream of dialogue and commentary, initiating reward sequences that temporarily satisfy the pleasure centers of the brain, whether through a consumer purchase or promising social connection. What I find particularly challenging about these mutually reinforcing processes—internet surfing and consumer access—is how I waiver between a sense of personal autonomy (I control my journey) and subtle manipulation (I'm being guided by processes that I cannot see).

In *Irresistable*, Adam Alter suggests that the internet (and other technologies) intrinsically, and indeed nefariously, support behavioral addiction, and various media companies and outlets are aware of this. They develop techniques and algorithms that prey on that potential, hooking us into their products and programs, ultimately creating a nationwide and even global behavioral addiction syndrome. While we're focusing on the business of daily life, our brains are processing multitudes of information below the surface of conscious awareness. Meanwhile, this peripheral information subtly shapes our thoughts, feelings, and actions, and crafts some of our most critical life outcomes.[2]

I'm ambivalent about Alter's hypothesis. On the one hand, I feel that I've spent a lifetime cultivating autonomy and clarifying my intentions. Yet I also have a sneaking suspicion that he's right and forces beyond my control are subtly influencing my decisions. These forces may not be intentionally nefarious, although the recent Facebook controversies around propagandistic electioneering are sobering. But they are surely autonomic in the sense that the internet and computers predispose users to "know the world" in ways that are shaped by the perceptual impact of the technologies. Here is the ghost of Marshall McLuhan.[3]

Perhaps addiction is too powerful a metaphor. It assumes that a person has lost all control and is incapable of deflecting compulsive behaviors. Rather, we should consider how the integration of consumer capitalism and the internet promotes the seeds of addiction by way of compulsion, recognize that dynamic, and cultivate constructive alternatives that promote autonomy. This challenging task is intrinsic to human freedom as we strive to balance temptation with restraint, desire with moderation, and instant gratification with long-term fulfillment. The internet accelerates the speed of this challenge and places it before you as a prevailing dilemma.

The Anthropocene involves mutually reinforcing technological breakthroughs—the extractive capacity to transform natural resources while changing the material conditions of the planet, and the communications capacity to transform how information is gathered, collected, stored, and transmitted. These capacities, taken together, are splendid and impressive, transforming the planet, implying human grandeur, and at the same time summoning hubris and presumption, prompting ambivalence and uncertainty. Are we making autonomous decisions or are our lives controlled by forces beyond our agency? Unfold the layers of political controversy, and you'll find that this powerful question is at the source of considerable personal and cultural anxiety.

The Anthropocene casts a collective spell over human awareness. It's an autonomic process of global transformation. Our challenge is to understand how that spell influences our daily behavior as well as changes the way we perceive the natural world and communicate with each other. I am under no illusions. I am enchanted by the grandeur of this global transformation, despite spending a lifetime trying to cultivate a deeper awareness of both my experience of the natural world and my social relationships. If you even modestly enjoy the benefits of consumerism and the internet, then you are undoubtedly equally enchanted, and it's time for us to work together to break the spell. Otherwise the Anthropocene will forever blow your mind. How, then, do we reclaim autonomy in this time of global transformation?

By autonomy, I refer to enhancing the capacity for mindful deliberation, and cultivating a deeper understanding of the choices we make and the context in which we make them. Matthew Crawford, in *The World beyond Your Head*, suggests that we live in an age of distraction, constantly fending off direct demands for our attention, from the innocuous but ubiquitous public spaces used for advertising to the constant messaging of electronic media.[4] Unless you're living in isolation, it's nearly impossible to avoid this assault. Nevertheless, people do get on with their daily lives, and find ways to focus on what's meaningful, purposeful, and constructive. Somehow you have to figure out what matters most. The challenge is how to recognize this attention competition, understand how it subtly influences your choices, and then create behavioral cues that support constructive autonomy.

Ubiquitous Novelty

"Which is more interesting," Neelay asks, "Two hundred million square miles filled with a hundred kinds of biome and nine million species of living things? Or a handful of flashing colored pixels on a 2-D screen?"
—Richard Powers, *The Overstory*

The attention competition manifests in the demand for ubiquitous novelty. When you are electronically connected, unless you have constructed elaborate filters (more on that soon), you must submit to a proliferation of advertisements, alerts, scrolling headlines, tweets, location prompts, texts, and other vectors of instant messaging.

In these times, stories emerge fast and furiously. Whether it's the ephemeral news cycle, the latest gossip on the internet, or keeping up with the "instant" messaging of our friends and acquaintances, we live in a time of narrative proliferation. Which stories capture our attention? How do we balance the multiple realities we encounter? What is fake, and what is true? Who should we listen to, and what makes their narrative credible?

Spend a few hours watching the news and you'll observe a frantic competition to capture your attention. CNN uses the moniker "breaking news" for anything and everything. Every story is a scoop. Every insight is profound. Every pundit is an expert. CNN (like all contemporary news sources) is strategizing about how to claim your attention. And we've learned with Facebook, YouTube, and Twitter (courtesy of prominent politicians and entertainers, often indistinguishable) that with instant access to social media, anyone can be a news source. Credibility matters less than messaging savvy.

Go to a modern sports arena. Every moment is filled with blaring screens, contests, and cameos of the crowd, all competing for attention, and delighted when they have their moment of fame. Perhaps I am old fashioned. I remember attending baseball games in the 1950s and 1960s when there was quiet between innings. You could gaze at the grass and look at the sky. No longer. There is never a moment of peace. The giant screens fill the gaps between innings by showing us all waving to each other. Yes, it's all in good fun, but it's another example of the pressing demand for attention, mainly focused on a form of collective human narcissism.

Walk down the streets of any city, and you'll notice that most people can't go from one place to another without either a "smartphone" in their hand and/or earplugs in their head. The next time you're in a public place, use the opportunity to observe this. See how long you can go without plugging in, checking your phone, or messaging your friend. It may be harder than you think.

This attention competition, considered in its entirety as the interaction of consumer capitalism, social media, and internet connectivity, promotes a demand for ubiquitous novelty, further fueled by the expansion of global markets. Attention competition promotes an anxious urgency. This is the source of a profound tension. On the one hand, you have instant access to data and information, whether it's the best price, quickest route, or global news. And you expect instant delivery of whatever you require or request. Under these circumstances, the past and future have less influence on your day-to-day decisions. What's the value of experience, legacy, posterity, and tradition when you are immersed in the demands and necessities of the present? How can you think about a long-term future—let's say the well-being of the planet one hundred years from now—when you're fending off this intense and immediate competition for your attention?

Reclaiming Attention

Attention is the thing that is most one's own: in the normal course of things, we choose what to pay attention to, and in a very real sense this determines what is real for us; what is actually present to our consciousness. Appropriations of our attention are then an especially intimate matter.
—Matthew Crawford, *The World beyond Your Head*

I was teaching my five-year-old granddaughter (J) the concept of fractions by showing her the various ways to think about her sixth birthday. We used a lineup of twelve stuffed animals to explore the notion of thirds. When she successfully internalized the concept, we figured out the exact date at which she would turn five and a third. Surely this was a cause for celebration, so we headed off to the local (Keene, NH) Target so she could choose a gift on the occasion of turning five and a third. I must confess that I had never before set foot in a Target store.

J had previously visited a Target in Seattle. So she took my hand and led me through the aisles. She demonstrated amazing mapping skills as she knew exactly where to go, down to the very aisle, and pointed to her proposed gift. She knew exactly what she wanted. She had seen it at the Target in Seattle, and the store layouts were sufficiently similar that she had no problem locating the toy. It was a windup cat, with a chip that looped various purrs and meows. What she found most attractive was that the cat would poop on command. I was astonished at how she navigated the aisles and was able to avoid all the distractions along the way. Among the countless choices of toys, J knew exactly where to go and how to get there.

I was impressed with J's ability to avoid distractions and accomplish her goal. I admired her cognitive mapping skills too. Yet I was concerned that her focus revealed a form of compulsive behavior, triggered by a consumer cue lodged in her five-year-old mind, supporting a myopic direction in a world of too many choices. How did this pooping cat claim her attention?

We returned home shortly thereafter, and following a few hours of imaginary play with the pooping cat, the gift eventually lost its luster, confined ultimately to the dustbin of glitzy, unwanted toys. J would rather play with her expanding menagerie of stuffed animals, as she embeds them with personalities, stories, and narratives of her own making. When she creates scenes with her stuffed animals, she is reclaiming attention, using her imagination, demonstrating autonomy and independence, and developing play worlds of her own making.

Several weeks later, J and I were sitting on a couch, both of us immersed in our tablets. I was checking my email. J was playing a cartoonish game. She was manipulating the icons with great speed and considerable skill. She was utterly mesmerized. I looked more closely at what she was doing. I realized she had virtually no understanding of the content of the game, but did understand its reward structure. No one taught her how to use the tablet or how to play the game. She figured it all out on her own. She knew how to find her way around the screen, how the various scenes and icons connected to each other, and how to move from one virtual place to another. The content was colorful but utterly irrelevant.

J's mother (my daughter) is a third-grade schoolteacher in Seattle. She thinks carefully about learning, limits the amount of time J is exposed to screens, and makes sure she has unstructured time in outdoor settings. The family takes hikes, has a garden, and emphasizes direct social interaction. My daughter brings the same approach to her classroom. She hopes that this enlightened and reflective supervision mitigates the compelling power of screens, and thinks carefully about her role as both teacher and parent in that regard. So she puts limits on how much time J spends with the tablet. I have no doubt that if J had not been interrupted via parental supervision, she would have continued with the game for quite some time. Only her natural inclination to use her body and exude her kinetic energy would prompt a withdrawal.

The best way for J to break the spell and reclaim her attention is to bring her to a park or playground. There she experiments with her body, encounters other children and adults, explores interesting new spaces, investigates the relationship between built and natural structures, and does so unencumbered by the trappings of screens. When I observe her in these settings, I notice how free she feels. I vicariously experience her joy. I feel similarly when I take a long bicycle ride, hike in the hills, or meander on a city street. We are in our bodies and out of our heads, or at least we've created a healthy balance.

May I respectfully suggest that although J was only five years old when all this happened, her experience, the relentless demand for her attention, sparked by her exposure to the convergence of the consumer world and internet communication, is relevant to just about any person who has a cell phone. There are 6.8 billion cell phone subscriptions on the planet. If you've got one and you use it, you're ensconced in the human empire with all its trappings. May I also suggest that these competing demands for attention are not just the province of privilege, although access to money certainly changes the rules of the game. We live in a world of proliferating choices, and if you don't have the means to attain them, you may well wish that you did.

Is Autonomy Possible?

How can the insights we already have in hand be used to foster the best of our behaviors and lessen the worst?
—Robert Sapolsky, *Behave*

The first step in reclaiming attention is to determine where it is you want to go. This seems like the most basic precondition for clarifying intention. You understand the choices before you, explore the alternatives, make a decision, and then proceed. I'll propose (soon, I promise) how focusing your attention on your ecological place can serve as a center for this practice. Before doing so, I'll briefly elaborate on some of the obstacles we inevitably encounter.

Clarifying intention presumes you're capable of rational, deliberate behavior. In the last several decades of research in fields including evolutionary psychology, neuroscience, cognitive behavior, and all shades in between, we've learned that clear decision making is quite difficult and made even more challenging by the prospect that you think you're making autonomous choices when in fact you're not. There's an impressive (and frequently overwhelming) spate of current literature that outlines some of these challenges and then promotes ways to overcome them.

Psychologist Daniel Kahneman in *Thinking, Fast and Slow* explains that human behavior is characterized by two modes of thinking. "System 1 operates automatically and quickly, with little or no effort and no sense of voluntary control. System 2 allocates attention to the effortful mental activities that demand it, including complex computations. The operations of System 2 are often associated with the subjective experience of agency, choice and concentration."[5] In elaborating this typology, he covers many of the typical fallacies to which we are all prone: confirmation bias, jumping to conclusions, using anecdotal evidence, and the various illusions that accompany certainty. In so doing, he asserts that better awareness of these two modes of thinking will allow us to make better judgments in conditions of uncertainty and promote decision making more in line with our original intentions.

In *Wired for Culture*, evolutionary biologist Mark Pagel explains that "increasingly cognitive science teaches us that our perceptions and memories are not just fallible; they are stories our brain concocts to prop up our egos, justify our decisions, and condone our actions." His assessment is that "an important role for self-deception might be to allow the brain to produce a narrative of everyday existence that is somehow constant, in the face of the gnawing internal conflicts over what to think and how to behave."[6] Pagel draws these conclusions in part from his study of the evolutionary origins of tribalism. He intends to promote insight by explaining how prone we are to tribal instincts, how those instincts are so often delusional, and how we overcome them by stressing trust, common values, and cooperative exchange.

Sapolsky, a neurobiologist, reminds us in *Behave* that "we are constantly being shaped by seemingly irrelevant stimuli, subliminal information, and internal forces we don't know a thing about." His exceptional book covers an enormous amount of ground, tracing the origins and context of human behavior. By explaining our worst behavioral tendencies, he sheds light on how we can overcome them constructively. "Many of our best moments of morality and compassion have roots far deeper and older than being mere products of human civilization."[7]

These works (much deeper than I can possibly explore here) help us understand the conceptual incongruities of human social behavior. We may not act as autonomously as we'd prefer, but a better understanding of unconscious behavioral drives contributes to more reflective awareness of everyday decisions and moral outcomes.

This is just a small sample of some of the groundbreaking research that's emerging from the interdisciplinary convergence of evolutionary biology, neuropsychology, anthropology, sociology, and history. I cite these examples merely to suggest that the origins and manifestations of human behavior are still mysterious, we are subject to forces we don't understand, many of these emerging concepts are limited to scientific frameworks (as illuminating as they might be), and we should be humble when asserting our confidence in autonomy. Kahneman, Pagel, and Sapolsky, among hundreds of other researchers, also present approaches for helping us understand how to cope with these challenges. We are just beginning to understand how to apply their work and use it in educational settings.

Clarifying Intention

Where are we? I started the chapter by exploring my immersion in screens (shorthand for computers, smartphones, and the internet), my ambivalence regarding them, and how I use ecological place to reorient myself. I explained how I interpret the Anthropocene as the convergence of accelerating natural resource extraction and global communication networks, and how the dynamic pace of these processes has profound implications for how the psyche perceives the biosphere. I suggested that these processes, combined with global capitalist expansion, cast a collective spell, and in so doing, create proliferating demands for our immediate attention, yielding a quest for ubiquitous novelty. I reiterated my concern that these demands create a widening gap between our psyches and the biosphere. I complicated matters by referring to an emerging body of research that challenges conventional notions of autonomy by emphasizing that human behavior is formed by impulses we don't yet really understand. And now I contend that the Anthropocene brings all these dynamics to the foreground. The same processes that can seem so bewildering and demanding are also stimulating and encouraging. No one said that expanding human knowledge would be easy! So it shouldn't be surprising if the Anthropocene is blowing your mind.

What is one to do? How do we reclaim our attention? How do we break the collective spell? This is a book about environmental learning as a foundation for how we come to know the world. I believe that a more intimate awareness of natural history, the biosphere, and what I describe as ecological place grounds our attention in the here and now, allows us to create deliberate pauses in our lives, cultivates a broader understanding of what it means to be human, helps us construct meaningful lives, and contributes to human flourishing while respecting the profound legacy of ecological and evolutionary patterns. That is precisely why environmental learning matters.

Here are four predispositions that inform my thinking. First, clarifying intention is crucial to environmental learning. The attention competition almost always excludes any ecological context, and the best antidote is to pay closer attention to the natural history of the biosphere and all the different ways that's possible. I'll further develop this point throughout the book. Second, most people experience the Anthropocene by daily life in the world—how they interact with technology along with how they work, play, and survive. They don't use the word or consider its consequences. But they wonder how much control they have over their lives, how they can deal with the anxiety that often accompanies a perceived lack of control, and search for ways to reclaim control and autonomy whenever possible. Third, clarifying intention should be an educational mandate. Learners across the entire developmental spectrum are exposed to attention competitions, sometimes perceived as a lack of control, and learning how to cope with these conditions requires skill and experience. Fourth, clarifying intention is a collective challenge. Mindful deliberation is surely enhanced with personal reflection, but it does no good if you are entirely in your own head. Clarifying intention should be a public conversation as it inevitably leads to diverse ethical positions that are best explored through civic engagement. If the Anthropocene is blowing your mind, you won't find solace through self-help. There's a lot to figure out and many experiences to compare, and we need all the help we can get.

The remainder of this chapter presents some possibilities for clarifying intention, reclaiming attention, and using ecological place as the venue for doing so. I'll start by going back to basics, and that leads us directly to Henry David Thoreau.

How Can Thoreau Help?

Thoreau could speculate that even a slight shift in natural processes—a little colder winter, a little higher flood—might put an end to humanity, so dependent are we on a wild nature that gives us no guarantees. Hence he emphasized living "deliberately," that is, living, so as to perceive the moral consequences of our choices.
—Laura Dassow Walls, *Henry David Thoreau*

As Walls writes in her exceptional biography, Thoreau was witness to the first stages of the Anthropocene. He was an astute observer of the transformation of the New England landscape. He understood the impact of the railroad, implications of westward expansion, and profound ecological changes that would soon be unleashed. Thoreau was no Luddite. He appreciated the promise of technology, but was wary of its unfettered application. There was no internet then, but the telegraph and railroad signified a new communications regime. The pace of change seemed swift enough, and Thoreau's response was prescient.

He literally put his stake in the ground by devoting his life to the most detailed possible understanding of the landscape where he lived and writing about his impressions, while turning them into a moral manifesto. His time at Walden Pond was a magnificent educational experiment, and an opportunity to live close to the land while still having access to town. Thoreau was constantly testing his perceptual limits by both closely observing the natural history of the local landscape and deepening his conceptual understanding of the wild within himself as well as in the mystery of nonhuman nature. He explored the transition between his body and the world outside. He believed that he could come to know himself by closely observing natural history.

Thoreau fully understood the idea of proliferation. He was a natural history polymath, interested in every living thing, every landscape, and the past conditions that led to present-day, local ecology. The educational experiment of Walden Pond was his attempt to clarify intention and reclaim his attention in the midst of technological change. He did so by choosing to live simply, focus on what really mattered, avoid the trappings of an exclusively human world, see himself as a human organism, and make choices based on his immersion in place. This is how he came to know the world.

Here's the lesson for our times. Thoreau chose to live simply so he could focus on his priorities, not to narrow his choices or simplify his thinking. Rather, living simply would allow him to penetrate more deeply into the world outside himself, the natural world, and in so doing, he would better understand his relationship to nature, what I describe in an earlier work as ecological identity. He examined the extraordinary complexity and intricacy of ecological connections. He understood that he could only do so by removing unnecessary distractions or living simply. A simple life leads to more focused thinking. This insight is hardly unique to Thoreau, as it is the basis for the world's meditative traditions. But Thoreau's great contribution was to demonstrate how this meditative, observational, simple living approach was also the foundation for environmental learning.

Here's why. Thoreau understood that he was living in the early years of an unprecedented expansion of human influence on the planet. To better know himself, he wanted to better understand nonhuman nature by paying close attention to all living things. How is your place in the landscape influenced by topography, physiography, and ecology? Are you wary of living a life that is solely determined by human affairs? The best way for humans to understand the meaning of humanity is to look outside themselves, observe other species, place themselves in an ecological context, and imagine themselves in evolutionary time. The core of Thoreau's insight is that the best way to know yourself is to closely observe the natural world. His great concern was that a human-dominated planet would make such a task increasingly difficult. You can't know the world unless you expand beyond the human domain and investigate what else is out there.

Thoreau's essays take us out of our head and into the natural world. Our challenge is to make space for these experiences, make them a priority, and reclaim our identities as natural organisms in relationship to the complex ecological/evolutionary layers of our heritage. This is the best antidote to an exclusively human world and the rabbit hole of collective narcissism that's constantly demanding our attention. How can Thoreau help us navigate the Anthropocene? How do we clarify intention and reclaim attention? You start by structuring behaviors that take you, to use Crawford's expression, to the "world beyond your head," the ecological origins of daily life, and the very essence of what it means to be human.

Reading the Day

At a young age, I learned to love the newspaper. I started with the sports page. It connected me to baseball games played in distant cities. I then migrated to the international news. As I got older, I explored all the other features of the paper. I learned how to scan the paper for stories of interest and read between the lines. I began to understand that what isn't reported is also important. The daily newspaper taught me how to browse, surf, synthesize, and interpret information.

Sixty years later, I still wander through the morning news. But now I do so by visiting various websites or discussion forums. The speed and access of my approach is greatly accelerated, but the basic principles of browsing, synthesizing, and curating are strikingly similar to the habits I learned reading the daily newspaper as a child. I find that gathering information this way at the start of the day helps me get my bearings. It helps me focus my attention on what's important and what I hope to accomplish, even if I only have a few minutes to do so. Yet I find that beginning the day this way also predisposes me to the pace of the internet. It speeds up my thinking. I enjoy the routine I've established, but I know I've circumscribed my attention by spending these early moments in front of a screen. Even though I know better, I often can't help myself.

There's another great way to start the day, and that is to observe the biosphere. In his delightful book *Minding the Earth*, Joseph Meeker suggests that "it is too much to say that you are what you read in the morning, but it is a sure bet that you aren't what you don't. . . . It is worthwhile to pause for a moment and reflect upon the character of the Morning Reading pursued by each of us. . . . A good day in the life of a living system begins with recognition and affirmation of life."[8] Meeker goes on to reiterate the necessity of observational ecology as an educational practice.

The "Morning Reading" is an indispensable foundation for reclaiming attention to promote environmental learning. Consider two different ways of greeting the day. You can step outdoors wherever you may be in order to feel the temperature, wind conditions, light, sounds and smells, or whatever visceral impressions fill your senses. Or you can immediately glance at your phone to check your messages, email, or whatever virtual information that gets you oriented. These are different information-gathering habits or approaches to learning. They are not mutually exclusive. As electronic information comes to dominate so many learning landscapes, it's instructive to consider the differences and find ways of balancing how we greet the day. Whether you're attending to your children, getting ready for work, checking your email, or sipping your morning coffee, there might be someway you can add "reading the day" to your routine. Notice how it changes your pace, attention, and quality of life, especially when you experience stressful circumstances.

Passing Moments of Environmental Learning

Time for a walk
  In the world outside
    And a look at who I am
Originally I had no cares
  And I am seeking nothing special
Even for my guests
  I have nothing to offer
Except these white stones
  And this clear
    Spring water
—Muso Soseki, *Sun at Midnight*

If you can take the time for reading the day, why not structure more frequent deliberate pauses, moments of environmental learning, punctuating your daily experience, performed in the spirit of mindfulness, with no agenda other than reclaiming your attention, and doing so by focusing on your ecological place.

Just consider the idea of ecological place—human activities, cultural processes, built structures, natural landscapes, the trees and grass, the critters, the microorganisms, the flow of water, and the atmosphere. How do these elements connect and integrate your perceptual awareness? How do they lend meaning and purpose to your day? How do they influence your place in the community, presence on the planet, daily moods, and understanding of life?

Of course, these are challenging questions. Simply let them provide a context for your experience, even for short moments of time. Sixty seconds is all it can take. Engage in brief journeys. Modulate your pace. Do it whenever time allows. There's always an opportunity for a deliberate pause.

Part IV of this book explores these possibilities in more depth. Here are a few suggestions to get started.

- Notice the interface between indoors and outdoors when you leave a building or step out of a vehicle.
- Look away from your screen and refocus your gaze on the biggest perceptual field you can find.
- During a commercial or break when you are being entertained, mute the sound, close your eyes, and focus on the acoustics of the room you're in.
- As you walk from one place to another, leave your planning behind and pay attention to every nonhuman living thing.

Develop your own variations for coming out of your head and into the biosphere. Find the practices that work best, help you perceive the world anew, sprout your imagination, and stimulate further inquiry into the relationship between your daily awareness and ecological place. These are small steps that can change how you think, what you pay attention to, and how you reflect on your perceptual capacity. Awaken your senses, alter your pace, and enhance the moment.

Our challenge is to penetrate the layers of innocuous cues that inform behavior. You start by making a conscious effort to scrutinize the use of screens and internet access, proceed with mindful deliberation, be wary of compulsive behaviors, develop counter cues that signal danger, and develop behavioral reminders to return to your original intention, all to serve the enhancement of environmental learning. The behavioral reminders can be any of the infinite possibilities of the biosphere—the sky above your head, soil beneath your feet, crow on the roof, fly that just landed on your hand, or human organisms walking down the streets of your city.

This is both an educational challenge and daily practice. In educational settings, I advocate that teachers provide students with the skills and activities at the earliest ages to help them anticipate as well as understand compulsive behaviors, and supply them with behavioral cues from the natural world as a focus for attention. These skills are best formulated in group structures when students can work together to support their efforts, exchange cues, coach each other, and discover as a community how to reclaim their attention in the service of environmental learning.

Frequently it's the simple activities that are most effective. Learn how to look away from screens on a regular basis. Use your experience to determine the most effective environmental cues. Try to turn your attention to the natural world at fifteen-minute intervals. Learn how to balance all the demands on your attention. Learn how to see the world through different frames. Compare your observations in various circumstances. These are practices, relevant for citizens and students alike, that can help break the spell of the screen. They are surprisingly hard to maintain, but essential to consider.

Proliferation and Filters

The observer uses a filter to engage the world.
—Timothy Allen and Thomas Hoekstra, *Toward a Unified Ecology*

When you're engaged in a particularly important or complicated intellectual task, or when you simply want to savor an experience or a conversation, it's best to turn the information faucet down to a trickle.
—Nicholas Carr, "What Scientific Concept Would Improve Everyone's Cognitive Toolkit?"

In my preteen years, I had two mind-boggling optical instruments: a prism and teleidoscope. I was mystified by how these instruments could expand, amplify, rearrange, and transform my visual perception. The prism would take a simple light beam and turn it into a magnificent spectrum. The teleidoscope would turn any object into layers of unfolding symmetrical patterns. I would walk around the house with the prism or teleidoscope held up to my eyes, amazed at how my ordinary reality could be so easily transformed. I learned that any scene or object always contained much more than my first impression would surmise.

This was an early experience of proliferation, the process by which any idea or concept can rapidly reproduce into infinite branches of possibility. This type of conceptualization is an important aspect of human cultural experience. There are so many ways to engage in a proliferation experience—a walk in a crowded city, stroll in a forest, visit to a museum exhibit, or meander down the aisles of a library or bookshop. Many cultures used psychedelic mushrooms and herbs to facilitate powerful proliferation experiences, transforming the senses beyond their ordinary reality. In some respects, the internet is a similar proliferation experience, although it is narrowly bounded by its two-dimensional frame. Proliferation is a key learning tool as it can demonstrate the power of possibility while displaying what Darwin described as "endless forms most beautiful."[9]

But proliferation can also overwhelm you, and unless you have guidance, it can lead to endless forms of distraction or experiences you may not be able to handle. It takes discipline, practice, and skill to navigate potentially complex learning spaces. I wonder about the extent to which unsupervised social media and internet exploration, especially among youths, is like a negative drug experience: too much information too quickly. I wonder, too, whether social media consumers, somewhat daunted by the possibilities of too much exploration, find solace and safety in chat rooms of like-minded people—a self-induced safety and conformity measure. I'll cover this possibility in the next chapter.

How, then, do you balance the creative possibilities of guided proliferation with the prospect of dazzling confusion? That's where filters come in. In ecological systems, organisms constantly filter the extraordinary range of sensory inputs. That's a crucial neurological function, referred to as sensory gating, a physiological process or information filtering in complex systems. Learning how to construct cognitive filters is an important way to exercise relative autonomy in complex information environments. CEOs use executive assistants. Email programs allow their users to employ filters. Keyboard synthesis uses filters to create specific tone colors. Of course, filters also cut things out, possibly preventing you from perceiving the full picture.

I suggest that learning how to construct filters is an educational imperative. If the filter is too powerful, you will have a limited view. Without any filters, you may see more than you are capable of internalizing. Totalitarian states use censors to filter information that may prompt dissent or critique. Overly permissive systems often have insufficient limits.

The deliberate pause is a filtering tool to promote ecological perception, modulate pace, and promote mindfulness of the ecological present. We need more practice, as individuals and communities as well as within institutions, especially schools, in understanding how to construct effective filters that synchronize intention with attention, balance proliferation and pause, and enhance awareness without blowing your mind.

Collect and Consume

The Museum's 80 million specimens form the world's most important natural history collection. The scientific community is using the collection to answer key questions about the past, present and future of the solar system, the geology of our planet and life on Earth.
—British Natural History Museum website

But taxonomy is not just another science, born out of naked reason and using elegant experimentation to make its way steadily forward. Taxonomy is instead a science born out of an ancient human practice—the ordering and naming of life—out of the urgings of the human unwelt.
—Carol Kaesuk Yoon, *Naming Nature*

In the mid-nineteenth century, shortly after the Industrial Revolution, with the advent of European expansion and possibilities of more rapid oceanic shipping, accompanying what Rosalind Williams describes as the "triumph of human empire," a nascent middle class grew its disposable income. Before the mass production of goods, some of the most appealing consumer items in what Frank Trentmann refers to as the emerging "empire of things" were artifacts of natural history, especially those procured in distant corners of the world. Following a European tradition, originally among the elite, of wealthy households constructing cabinets of curiosities for the express purpose of displaying and contemplating wonder, such home exhibits became the province of the middle class. The natural history collection was among the most popular ways to fill these cabinets. Indeed, Alfred Russell Wallace and Henry Walter Bates supported their scientific inquiries by exporting specimens from the far reaches of the globe. When one of Wallace's shipments disappeared in its transpacific journey, he was bereft as it was his sole source of income.

Almost two centuries later, natural history collections as tools of wonder are more arcane, or merely the esoteric or even eccentric pursuit of nature freaks and environmental educators. The dark side of this pursuit is the search for animal parts (elephant tusks, for example) that supposedly endow special powers on their human recipients, but threaten species survival too. Collections can serve science, art, wonder, and curiosity, but they can also reflect a profligate life by supporting obsession, distraction, and hoarding. The most famous evidence was Imelda Marcos's private shoe closet.

Collections of consumer goods, artifacts, and experiences are surely a hallmark of modern life. And as Trentmann explains in his magisterial history of consumption, *The Empire of Things*, "In the rich world—and in the developing world increasingly, too—identities, politics, the economy and the environment are crucially shaped by what and how we consume. . . . How much and what to consume is one of the most urgent but also thorniest questions of our day."[10]

What we choose to consume and collect is a cultural manifestation of the Anthropocene, from global flea markets to mega box stores to amazon.com.

One environmental response to these unfettered trends is the quest to live simply. This was at the heart of Thoreau's journey, and as David Shi pointed out in *The Simple Life*, also a prevailing theme in US history. Whether it's an emphasis on craft, the desire to return to nature, or a periodic household purge, we are all trying to figure out when enough is enough.

The British Museum's eighty million specimens need to be stored in one place, not only so they can be appropriately studied, but to ensure that they don't disappear forever. So we also collect to both study and save.

In a world of proliferation, how do you make decisions about what to consume and where to tread? Which paths do you follow, and why do you take them? We are all curators now and need guidance to better understand the deeper meaning of curation. Why is curation so important as well as a foundation for environmental learning?

Curate and Connect

But I believe "to curate" finds ever wider application because of a feature of modern life impossible to ignore: the incredible proliferation of ideas, information, images, disciplinary knowledge, and material products we all witness today. Such proliferation makes the activities of filtering, enabling, synthesizing, framing, and remembering more and more important as basic navigational tools for twenty-first century life. These are the tasks of the curator, who is no longer understood simply as the person who fills the space with objects but also as the person who brings different cultural spheres into contact, invents new display features, and makes junctions that allow unexpected encounters and results.
—Hans Ulrich Obrist, quoted in *This Will Make You Smarter*

Here is the psychological dilemma that both inhibits and enhances environmental learning. We live in an era of unbridled cultural proliferation, fueled by the accelerating pace of human infrastructure and communication, tempted by the prospects of collection and consumption, representing the chimera of identity. We require filters, cultivated through our values, to sift through the possibilities, set priorities, and forge learning paths that promote human flourishing on a dynamic planet. We clarify intention by adhering to our values. We reclaim attention by constructing filters that establish priorities.

But filters leave out as much as they let in. We have all encountered narrow-minded people who are stuck in limited frameworks that prevent learning and growth, exclude more than they include, and circumscribe the boundaries of their lives. Sufficiently reflective people will be open enough to recognize such behavior in themselves.

Both biologically and psychologically, life is a constant tug between proliferation (opening yourself to exciting sensory and learning opportunities) and filters (knowing what to keep out and what to let in). If the Anthropocene is blowing your mind, this is what you must manage.

The word "curate," originally reserved for art galleries and museums, is now widely used in many settings. We now have social media curators who synthesize and interpret information, or as defined in a *Huffington Post* article, spend their time "filtering through all the interesting content across the web and sharing the best news, articles, videos and infographics on your social channels."[11] A 2017 *Forbes* article, "Your Top Seven Tools for Social Media Curation," lists various computer programs that help you "find relevant, interesting, and valuable content to share with your audience."[12] There are thousands of bloggers who have used this skill—merged with interpretative orientations based on the interests of their readers—raised money through subscriptions, and developed robust audiences of "followers." Indeed, I have a section on my bookmarks bar labeled "idea blogs" so I can curate for myself the curators I find most helpful.

The process of curating the curators is sometimes more helpful than actually following their blogs. When I curate the curators, I clarify my intentions. What is it that interests me? What will enrich my thinking? What will distract me? Apparently when I was a young child, I used to enjoy taking books out of bookshelves and then putting them back again. Nearly seventy years later, I still do the same thing, constantly shuffling and reorganizing, taking ideas apart and putting them back together again, both virtually and viscerally, on the internet, and then with my home library, the music I listen to, the board games I play, and the walks I take. What direction shall I take? What do I wish to learn about? How shall I spend my time?

While I fully understand that the choices I make may not fully be under my control, my aspiration is to reflectively consider those choices by linking them to what most matters in my life. This book is about why environmental learning matters, so our challenge, as learners and citizens alike, both in private and public, is how best to curate environmental learning.

Obrist's advice above is profound. He urges the curator to bring different cultural spheres into contact while making connections that allow unexpected results and encounters, or to use a colloquialism, to take you out of your comfort zone. I will address these challenges in more detail later in the book when I discuss improvisation.

For now, I wish to stress the necessity of designing learning experiences that reflect your values-based intentions and allow you to focus attention in ways of your own choosing, with an emphasis on promoting filters that modulate your use of screens and proliferate your Thoreauvian impulses. This is the perennial challenge for the environmental educator, made even more pertinent by the accelerating pace of distraction. Use your deliberate pauses well, and use them to read the day so your ecological place is a primary focus. Organize your consumption and curate your collections to sharpen this focus.

How you curate your learning is much more than an individual challenge. It's a social process, linked to peers, friends, family, and neighbors, your most worthwhile and vital connections. In the internet era, when social connections take such primary importance, what networks will you participate in, and how will they enhance your environmental learning? Connectivity is the hallmark of our times. But how do we promote constructive connectivity—networks that simultaneously promote human flourishing and ecological place, while ensuring cultural and biological diversity?

III
A New Vision for Environmental Learning

5
Constructive Connectivity

Visiting the Golden Spike

The railroads, like that signal across the wires, caused millions to imagine events and possibilities they could neither see nor hear. The transcontinentals were from the beginning always running ahead of schedule, always approaching places that did not quite exist. The transcontinentals were the means to something beyond themselves even if there was a certain lack of clarity about what those things might be.
—Richard White, *Railroaded*

Scientists in the seventeenth and eighteenth centuries discerned that there were networks in nature—from spiders' webs to the human circulatory system of veins and arteries—but it was not until the nineteenth century that the term began to be used more metaphorically, by geographers and engineers to describe waterways and railways, and by writers to characterize the relations between people.
—Niall Ferguson, *The Square and the Tower*

The golden spike, the symbolic rail fastener, was the final connection for the Transcontinental Railroad, linking the US East and West. You can visit the national historic site in Promontory, Utah, and I did so, complete with its reenactments, a gift shop with railroad books, and a few trails through the desert. The vast, unsettled, arid landscape belies the connectivity of this historic network milestone. Today it's a relic. As you hike through the surrounding desert and gaze at the pilings, you can reimagine the thousands of hours of excruciating labor (mainly performed by Chinese migrants), investment of financial capital (mainly organized through slippery loan schemes), and upheaval of the surrounding landscape (boomtowns, dislocations of indigenous peoples, and places left behind).

As White writes, "The transcontinentals opened the question of a national market and the relation of the East and the West."[1] A journey

Figure 5.1
The golden spike

that once took six months would now take a week. Commerce could flow unfettered from sea to sea, with various stops along the way, weaving regional transportation networks into a dynamic grid. The telegraph and railroads allowed messages and people to move at unprecedented speeds. But more important, they expanded their users understanding of networks along with how distances could be transcended, connections were both local and global, and mobility accelerated the movement of people, goods, and capital.

In England, as Janet Browne explains in her two-volume biography of Darwin, the "length of railway track snaking across the countryside doubled from 1850–1868." This enabled the proliferation (there's that word again) of correspondence. "By mid-century, 600 million letters were dispatched every year. Twenty-five thousand delivery men travelled 149,000 miles to distribute these letters, carrying in their sacks a weight of nearly 4,300 tons." This included newspapers, book parcels, money orders, and letters. As Browne notes, "If there was any single factor that characterized the heart of Darwin's scientific undertaking it was this systematic use of correspondence."[2]

Remember from the previous chapter that some theorists proclaim that the invention of the steam engine (1784) launched the Anthropocene. Without the steam engine, there are no railroads, steamships, or timely long-distance mail connections. Darwin's correspondences had a significant carbon footprint. Nineteenth-century European expansion, fueled by the steam engine, organized to dominate markets and colonize natural resource locations, also promoted voyages of mapping and scientific discovery. The theory of evolution was ultimately catalyzed by the ability of scientists to share data over landscapes and oceans. Networks move both goods and ideas.

Over the course of the next one hundred years and into the present, the "triumph of human empire" has been the triumph of visible network infrastructures—telephone lines, television antennas, electric power grids, and interstate highway systems—and those that are buried or invisible too—transatlantic cables, radio waves, and the internet.[3] Consider that these have mainly been built over the last one hundred years. Imagine the ecological impact of these structures, including the energy required to build and power them.

Networks are essentially venues for moving things from one place to another, or communicating signals and data. Hence they are intrinsic to organisms and ecosystems. They form splendid patterns, depicting ebbs and flows, branching and fractals, circulations and convergences. They are as ancient as life, manifested in cellular signaling systems, the assemblages of mycelia, and root systems of trees. They are fundamental to human culture, manifesting in trade routes, social relationships, and transportation processes.

Despite the ubiquity of networks in nature and culture, how they work and how to utilize them is not self-evident. Humans have always used networks to accomplish their goals, but with new technologies of visualization, and the increasing complexity and intricacy of human networks, they are the subject of copious interdisciplinary research. Indeed, the ability to visualize, construct, and maximize networks is profoundly important for professional and personal success.

The Transcontinental Railroad and telegraph were nineteenth-century road and communication systems that imprinted a conceptual template on the people and cultures that used them. With the invention of electricity and computers, the complexity of human networks takes on a new level of dimensionality, sprouting connections faster than they can be comprehended. The railroad no longer travels through Promontory. In fact, cross-country rail travel is now a US anachronism, a romantic jaunt for old-fashioned passengers. When I visited Promontory, I imagined the ghosts of railroads. To learn more about the history of the golden spike, I could use my cell phone.[4] I noticed four bars of reception! I could connect to the world quite easily from a place much more notable now for its splendid isolation. Still, there might be no internet if the railroad and telegraph hadn't preceded it.

One hundred and fifty years later, connectivity is much more involved and every bit as engaging. Its prospects can be unifying and polarizing. It allows for exceptional transmission speeds and instantaneous feedback, changing the very protocols and culture of human communication. Connectivity can also lead to overload or the tyranny of like-mindedness. How do we ensure forms of constructive connectivity—approaches that bring people together as well as facilitate deeper understandings of both diverse cultures and the natural world? How can network thinking better inform environmental learning?

Please consider this chapter as a participatory journey or an educational field guide for thinking about networks. I'll start with a network observation interlude. What networks engage, surround, and embody your world? What is constructive connectivity, and why is it a priority for environmental learning? I'll explore biosphere network archetypes by introducing the network structures that embed ecosystems. I will then briefly cover some of the latest research about ecological networks as well as some innovative new approaches to network visualization. From there I'll move to social networks, including some guidelines for intentional network navigation, and conclude by revisiting the concept of constructive connectivity.

Network Observation Interlude

Something marvelous is happening underground, something we're just learning how to see. Mats of mycorrhizal cabling link trees into gigantic, smart

communities spread across hundreds of acres. Together they form vast trading networks of goods, service, and information.
—Richard Powers, *The Overstory*

Thoughts and emotions that create bonds of attachment between us have no difficulty in crossing seas and decades, sometimes even centuries, tied to thin sheets of paper or dancing between the microchips of a computer. We are part of a network that goes far beyond the few days of our lives and the few square meters that we tread.
—Carlo Rovelli, *The Order of Time*

A network is a series of intricate pathways that facilitate the flow of information, energy, and matter. At the most basic level, a network is simply a line that connects two nodes or a two-way exchange. It's easiest to conceive the most basic network structure as a dyadic relationship between two people. Once the relationships expand to include multiple nodes (or many people and species), they get incredibly and wonderfully complex, beyond the scope of everyday cognition, operating at multiple scales. Whether you're thinking about your network of social relationships, the subway system under your city, or the mycelia in the forest, the expanse of networks is vast and often invisible.

To engage in "network thinking" means that you attempt to perceive the patterns of that flow. The patterns themselves are fascinating to observe, illuminating a variety of interesting topological structures. These structures will vary according to the type of network, scale at which it functions, and cognitive abilities of the perceiver, laying the groundwork for engaging interpretative possibilities. From an educational perspective, our challenge is to consider how we learn about networks, and then how that understanding promotes deeper learning. Let's engage in a deliberate pause and take some time to observe the networks in our lives, both visible and invisible, making note especially of all the networks that surround and influence us, typically just below the surface.

Notice first the visible network infrastructures that surround you. In each case, imagine the extent and expanse of the network, both in your immediate vicinity and then beyond. You might notice wires that plug into an electric grid. You have a network wiring scheme for your residence, connected through electric transmission lines to an energy source. Perform a Google Image search for electric power grids to better understand the extent of these networks. Electricity networks are frequently noticed when they stop working. When you lose electricity, you wonder where the network failed.

Old-school electric networks use wires, but we also have wireless, and in many localities a combination of both. When you are in a public space, your smartphone will list every wireless network within proximity, a bewildering cacophony of locked networks, connected through a common server, but only accessible to those who can unlock them.

Transportation networks—interstate highways, subways, railroads, airplanes, and bus routes—are visible at an accessible scale, although they are much easier to understand with mapping aides. Check out the app flightradar24.com to see all the airplanes in the air at this very moment. Telephones (both landlines and cellular) have intricate network routers and systems, stretching both around the planet and out into space with the use of satellites. Radios (if you still have one) are an example of wireless transmission and reception. Think about all the radio waves that are floating around the earth, bouncing off the ionosphere, moving through your body, and then coming into your radio receiver. The internet integrates all these functions—wires, wireless, sound, visuals, and mail—into a comprehensive communications package, available through devices that you can put in your pocket.

How about plumbing networks, the vast array of pipes, sewers, and waste that flows throughout towns and cities? Or the delivery system that brings water into homes and businesses? Or the movement and flow of money along with other economic currencies?

Think next about your social networks, the people you connect with. There are professional networks, organized around workplace functions, organizational charts, and institutional connections, further integrated through conferences and various professional associations. And there are your personal networks, links to families and friends, neighborhood and community, or hobbies and interests.

Finally, consider all the ecological networks that impact your life but that you rarely think about—cellular transport systems in your body, underground root systems, mycelia, food webs, biogeochemical cycles that flow through the biosphere, and migration routes of humans and other species. Later in the chapter, I'll discuss the importance of network visualization techniques in helping elucidate biosphere patterns and ecological processes.

An interesting way to understand and penetrate the depth of these networks is to draw them. I find this is a useful activity, and often apply it in both consulting and teaching venues. It's instructive for members of an organizational team to draw network maps (both professional and personal) to consider the reach of their efforts, who they influence and who influences them, and where they get support from, and then code the network diagram according to the strength of the various connections. When team members share their diagrams, they get an interesting overview of their collective influence and impact, and can use these data to further strategize their objectives. It's also an interesting way to trace the flow of power and influence in organizational life. In teaching settings, it helps learners understand where they get information from and all the subtle influences on how they think. Drawing ecological networks provides greater insight into their significance in sustaining life in the biosphere.

While drawing networks, it's particularly beneficial to discern structural patterns that help determine forms and venues of connection. It's interesting to compare visceral and virtual networks, networks that are obvious and those that are hidden, and finally determine which networks matter most. These are the skills that amplify and distinguish network thinking. I'll address them in more detail in the forthcoming sections on visualization and network navigation. For now, dwell in the space of considering the networks in your life, how you can better understand them, and how you might use them as a learning tool.

The central focus in this chapter is how to perceive connectivity among and between people and ecosystems. A reminder of our purpose is in order: how to promote a deeper understanding of human flourishing and ecosystem integrity—the motivating value that inspires environmental learning. By constructive connectivity, I refer to perceptual capabilities that enhance such learning. Perceiving connections is enlightening, but what's most important is determining the connections that matter along with the nature of those connections, and then bringing them to the forefront of learning.

In the following sections, I'll introduce some of the exciting new research in ecological and social networks, with an emphasis on how that research helps us develop criteria for constructive connectivity. But first let's consider the ancient networks of fungi, or what Paul Stamets describes as the "original Internet," as our path to many of the ways that networks are intrinsic to human activity in the biosphere.

Biosphere Network Archetypes

I believe that the structure of the Internet is simply an archetypal form, the inevitable consequence of a previously proven evolutionary model, which is also seen in the human brain: diagrams of computer networks bear resemblance to both mycelium and neurological arrays in the mammalian brain.
—Paul Stamets, *Mycelium Running*

Mycorrhizas form an infrastructure of interspecies connection, carrying information across the forest. They also have some of the characteristics of a highway system.
—Anna Lowenhaupt Tsing, *The Mushroom at the End of the World*

Mycelia are a networks of fine, white, branching filaments, the connective tissues that simultaneously decompose organic material while providing nutrients to plants, typically forming a symbiotic relationship, crucial to the health of a forest. Plants use mycelia to communicate with each other. As David George Haskell writes in *The Songs of Trees*, "Roots converse with fungi, sending chemical messages through the soil." These messages contain information "about attacking insects and drying soil." Haskell compares the soil to a street market where "roots gather to exchange food and in so doing they also hear the neighborhood news."[5]

These networks are extensive, stretching in some cases across several thousand acres. Fungi are the earth's decomposers, preparing the substrate on which all land-based life depends. Their reproductive spores, often appearing as mushrooms, are vectors of communication and transportation. When you see a mushroom, you are observing an ephemeral fruiting body of what is typically a vast underground network. Mycelia, active agents of forest exchange, information, and messaging, form a robust and efficient ecosystem exchange process, going back at least 400 million years and associated with the origin of land plants.

Compare the networks of mycelia, tree roots, watercourses, brain neurons, galaxies, and the migratory routes of the sooty shearwater.[6]

The sooty shearwater (figure 5.5) navigates most of the Pacific Ocean as it searches for breeding grounds and nutrients, flying forty thousand miles each year, cruising between hemispheres.

We perceive trees as single organisms grouped together. They nevertheless form underground networks and exchange nutrients among themselves (even among different tree species), and participate with mycelia in a comprehensive forest nutrient regime. Dendritic drainages, one of many geomorphological watercourse network patterns, are also nutrient and information transport systems. Animal migrations form complex webs across the planet. Brain neuron networks are an intricate organismic messaging system. Galactic networks are possibly communication systems too!

Yet these networks, fundamental to evolution, ecology, and species survival, are not typically visible at ordinary scales of human observation.

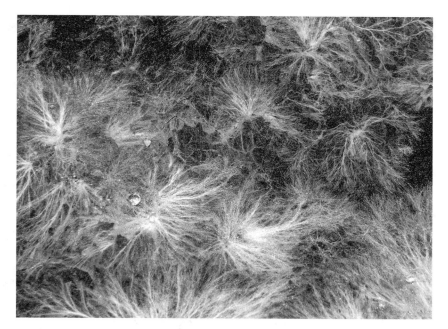

Figure 5.2
Agaricus bisporus mycelium. Photographed by Rob Hille (own work) [CC BY-SA 3.0 (https://creativecommons.org/licenses/by-sa/3.0)], via Wikimedia Commons.

Figure 5.3
Tree roots

Figure 5.4
Watercourses (dendritic drainage)

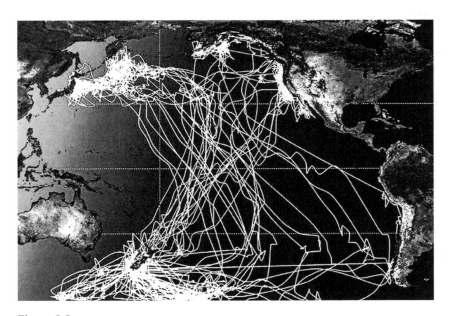

Figure 5.5
Sooty shearwater. Map courtesy of National Audubon Society Bird Guide Training Curriculum Basic Level.

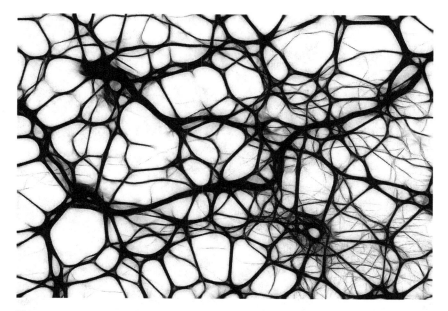

Figure 5.6
Brain neurons

Mycelia and underground tree roots are below the surface of the earth. And you can only observe drainage patterns from an airplane or mountaintop. The sooty shearwater migration path is best visualized with computer mapping techniques. Cellular networks like brain neurons are microscopic. To see galaxy networks, you need powerful telescopes supported by advanced computers.

I suggest these patterns are biosphere network archetypes. They are multiscalar, spatial configurations that reflect the dynamic movements of organisms and biosphere processes across space as well as time. I use the word "archetype" in the Jungian sense, as universal and archaic symbols that resonate through human history, derived from our collective observations of dwelling on this planet. They are inside and outside us, embedded in every layer of mind and consciousness—the very essence of life itself. With every breath, we connect to biosphere networks, although it takes mindfulness and study to illuminate these relationships. As we study these patterns in more depth, we deepen our understanding of life processes in the biosphere and hence they are a foundation of environmental learning. With computer graphics and the proliferation of environmental change data, we can now develop pattern representations of network phenomena in ecosystems.

Figure 5.7
Galaxies

Ecological Networks and Their Visualization

Ecological network theory is a rapidly growing approach to ecological research interpretation and synthesis. A report on the literature, "Ecological Network Metrics: Opportunities for Synthesis," defines network ecology as "the use of network models and analyses to investigate the structure, function, and evolution of ecological systems at many scales and levels of organization," and then describes the "influx of network thinking throughout ecology," citing a rise of relevant articles in the primary ecological literature from 1 percent in 1991 to 6 percent in 2017.[7] The article then lists examples of the many dimensions of ecological research that now apply network analysis. It's instructive to reproduce some of that list here, and I will do so without the citations attached to each topic: population dynamics and the spread of infectious diseases, food webs and nontrophic interactions, animal movement patterns, spatial studies of metapopulations, connecting biodiversity to ecosystem functioning, identifying keystone species, the sustainability of urban systems, and elements of the food-energy-water nexus.

Another comprehensive report, written by an international team of ecologists and published in a 2018 issue of *Population Ecology*, is "Complexity and Stability of Ecological Networks: A Review of the Theory."[8] The team's synthesis of the research includes a comprehensive taxonomy of network functions and assessments. For example, criteria to assess network complexity include species richness (total number of species in the network), connectance (proportion of realized interactions among all possible ones), connectivity (total number of interactions), linkage density (average number of links per species), and interaction strength (weight of an interaction in the interaction matrix), among others. All this leads to a detailed discussion of how to assess network stability and complexity.

These variables can boggle the mind. How do you portray them so a layperson can better understand ecological networks? Are there patterns to look for in assessing ecosystem integrity, and how can you illustrate them? There are many methodological challenges, especially how to synthesize the complexity of metrics into standard representational forms. Ultimately, ecological researchers are striving to understand the complexity of ecological dynamics and provide interpretative guidance about the connections that matter most. With advanced computer graphics, there's also a proliferation of visualization possibilities. Clear illustrative graphics of ecological networks can inspire as well as motivate scientists and laypersons alike. I'll say more about all this in the following sections on visualizing the biosphere and network navigation. For now let me simply endorse the prospect of two emergent fields linking insights: ecological network theory and information graphic design. And let me suggest further that these insights will be made immeasurably more helpful if coordinated with both graphic design artists and environmental educators.

Network concepts are also informing one of the fastest-growing environmental fields: urban ecology. In her groundbreaking book *Cities That Think like Planets*, Marina Alberti describes cities as hybrid ecosystems—that is, they are coupled human-natural systems, and as such, their "networks are not governed by either natural selection or human ingenuity alone" but rather they involve "new hybrid networks." Essentially, these are novel and largely unprecedented (in size, scope, and scale) combinations of ecological, human, and built networks, yielding entirely new ways of thinking about cities as ecological systems. Alberti's book explodes with interesting approaches to research and learning, but underlying it is her view that "understanding the topology of interactions among components is an essential step in decoding urban landscapes and relationships between patterns and functions," and that "network theory is emerging as a promising approach to uncovering general rules of complex landscapes."[9]

Let's get old school for a few moments. When Thoreau was roaming the hills and fields of greater Lincoln, MA, he did so without sophisticated network visualizations. Rather Thoreau, like all the people who wandered the planet for thousands of years before the Anthropocene, whether for survival or pleasure, engaged in direct, intense observation of the landscape, taking careful (written or mental) notes about the intricate lives and relationships of the great variety of species and habitats surrounding them. There are many ways to accumulate data. What matters is how the information is synthesized and then interpreted. Computer graphics can help, but none of this research is possible unless someone goes out into the field and gathers the data.

In his now-classic landscape ecology text, *Land Mosaics*, Richard T. T. Forman reveals how an understanding of spatial relationships is crucial in helping us interpret ecological management and restoration processes. There is a direct, mutually interactive correspondence between structure and flow. "Not only do flows create structure, but structure determines flows. For example, the arrangement of patches and corridors determines the movements of vertebrates, water, and humans across the land. Finally, movement and flows also change the landscape over time, much like turning a kaleidoscope to see different patterns."[10] Here is an ecologist's vision of the "girl with kaleidoscope eyes." Environmental learning requires that we see the biosphere with kaleidoscope eyes.

Network patterns can be observed at multiple scales, and they are manifest in the flows of matter, energy, people, and species anywhere you care to look. That is why networks are instantly familiar. I think that is also why they are aesthetically appealing. They link different scales of human observation to the grand patterns of the biosphere. The scientific community is growing its expertise in learning how to research, illustrate, synthesize, and interpret the connections in these networks. This research, to be meaningful, must simultaneously uncover complexity while delivering clarity. Where are the patterns, and what support do we need to better detect and understand them? How do we promote a deeper understanding of human-nature networks on a planet of over seven billion people along with an infinite variety of species and landscapes?

Visualizing the Biosphere

Each element within our collective biosphere will need to be carefully tracked, interrelating multiple factors such as geologic and climatic conditions, predatory traits, and life cycles. In this complex and multidimensional data environment, the role of visualization will be key in providing the capacity to recognize the emergent patterns and processes of these phenomena. Visualization will itself become organic, as it will need to adapt to simulate information from a wide spectrum of sources, ranging from micro/organic to macro/planetary states.
—Christopher Grant Kirwan

Burgeoning interest in these visualizations suggests that they may also prove useful for illuminating the most complex and important network of all: Earth's biosphere.
—David McCanville

These two passages, excerpted from a suite of essays at the conclusion of Manuel Lima's *Visual Complexity*, emphasize the necessity of using network visualization techniques to enhance and advance environmental learning. The challenge is clear: the conceptual complexity of the earth's biosphere is a daunting educational prospect. Hands-on natural history provides the observational skills, local data, and on-the-ground experience to inspire and inform environmental learning. Perceiving the biosphere may start in your local ecological community, but it requires an understanding of scale that transcends sensory observations and local knowledge. Perceiving networks is a conceptual skill that bridges the gap between local observation and biosphere-scale interpretation. Networks are biosphere archetypes intrinsic to the ecological messaging of every perceiving organism. Networks are familiar, intimate, and mysterious.

In the previous chapter, I asked how Thoreau can help. Now I ask how Lima can help. If you're not familiar with his work, you should be. His approach to visualization holds great promise for environmental learning because it blends a sophisticated awareness of computer graphic design with an understanding of human conceptual archetypal expression. His two-volume sequence, *The Book of Trees: Visualizing Branches of Knowledge* and *The Book of Circles: Visualizing Spheres of Knowledge*, when combined with his book about networks, *Visual Complexity: Mapping Patterns of Information*, provide magnificent displays of how to interpret complex ideas. In *Visual Complexity*, Lima brilliantly explores the interface between network science and information visualization, rightfully asserting that reflective network thinking is crucial as "our efforts at decoding complexity need to be mastered and consolidated so that we can contribute substantially to our long journey of deciphering an increasingly interconnected and interdependent world."[11]

Each volume develops a taxonomy of structures, located in the history of human visualization techniques. We find tree- and circle-oriented visualizations dating back to the origins of artistic illustration. You don't need a computer to draw a tree, circle, or network, but these archetypal forms multiply into an extraordinary variety of possibilities, many of which capture interesting ecological patterns. As the data get more complex, the patterns proliferate, but they still conform to an archetypal taxonomy. Trees and circles are also helpful visualizations for perceiving the biosphere. Trees involve intricate branching patterns. Circles depict loops and cycles. Networks, trees, and cycles are all contained within each other. Visualizing them promotes a deeper understanding of layers, hierarchies, flows, and relationships.

Lima's efforts, along with a copious emerging literature on information visualization, are impressive.[12] Most of this literature emphasizes psychological and aesthetic rules of thumb for enhancing as well as clarifying the graphic design challenge, although Katharine Harmon advocates the artistic use of maps for broader educational purposes. All these information design theorists remind us that our study of network taxonomies is emergent, embryonic, and evolving. They are the vanguard of an educational frontier, combining aesthetic design, network theory, computer graphics, and symbolic archetypes. What we need are ways of translating this literature into educational settings, and for the purposes of this project, using it to enhance environmental learning.

I would love to manipulate time to put Thoreau (or Darwin, Humboldt, or Rachel Carson) in the same space with Lima and his colleagues. I'd like to listen to a discussion as to how direct, hands-on natural history observations could be enhanced with these brilliant visualization possibilities. The next best thing is that we imagine how to integrate these approaches, and do so by experimenting with network visualization as informed by natural history observation and the urgency of global environmental change. Let students everywhere organize projects that blend these approaches to learning. Let's turn the internet into an encyclopedia of natural history knowledge enhanced by biosphere-scale visualizations. This would be an ideal blend of visceral and virtual approaches so that constructive connectivity incorporates network thinking on behalf of environmental learning.

Now let's turn our attention to social networks. Then I'll consider how to put all this together.

Social Networks

And not before 1980 was "network" used as a verb to connote purposive, career-oriented socializing.
—Niall Ferguson, *The Square and the Tower*

Social networks evolve from individuals interacting with one another but produce extended structures that they had not imagined and in fact cannot see.
—Charles Kadushin, *Understanding Social Networks*

As an experiment, when I first learned about LinkedIn, I thought it would be interesting to see how many connections I would get through utter passivity. I would happily connect with most people who asked to connect with me, but I wouldn't initiate any connections on my own. As of this writing, I have 673 connections—people who for one reason or another contacted me. When I peruse the list, I can make some educated guesses about why various people connected with me, but I only personally know about half the people in what LinkedIn describes as "my network." I have never used LinkedIn for any professional reason so I have no idea how and whether it works. If I do need something professionally, however, I am much more likely to think through my informal networks and get in touch with the appropriate person. These networks were established through experience, friendship, reciprocity, and a lifetime of sharing common goals and interests. I haven't taken the time to count the number of the connections that really matter in my life, and if I did I could probably divide them into various tiers and segments, thereby creating an interesting diagram. I'm sure the resulting structure would be instructive. What matters most is the social trust and reciprocity that is at the core of those relationships. I've worked hard to serve as a connector, and it gives me pleasure when I can bring people together to form a social or professional bond.

What are the social networks in your life, and how do you use them? There are many ways to organize an answer to this question. It's surprisingly helpful to visualize these relationships by mapping them. You can create a series of maps to describe kinship networks, associational networks (people you share interests with), professional networks, or any other designation that works for you. It's useful, too, to conjure these connections through contemplation and memory. Who comes to mind when you think about your social networks? Typically, the most salient and influential relationships are the people (or nodes) who connect various networks. Here's why.

Social network research, like other branches of network theory research, exploded in the last decade. This is an outcome of two converging trends: the increasing use of electronic media for two-way communication, and the expanded capacity to create virtual relationships. Facebook and LinkedIn are the most obvious contemporary venues for supporting (and creating) these trends. There's still much debate about the average size of any individual's network, but most research suggests that the circle of contacts in our lives is expanding.

Still, like-minded people tend to cluster together, finding each other in organized social settings or on the internet. The sociological term for this is "homophily," or the likelihood that any group of people with common attributes will have a social connection. In *Understanding Social Networks*, Kadushin succinctly summarizes the conditions of homophily: "In sum, if people flock together, it appears there are four processes involved: (1) the same kinds of people come together; (2) people influence one another and in the process become alike; (3) people can end up in the same place; (4) and once they are in the same place, the very place influences them to become alike."[13]

Social network analysis, like ecological network analysis, has an interesting and useful taxonomy of relationships, and it's instructive to explore them. In this section, I'd like to focus on just three terms: homophily, clusters, and weak ties. A "cluster" is simply a way to conceptualize the segmentation of networks. You and I belong to multiple networks. Are they separate or connected? And if they are connected, what is the nature of the connection? The term sociologists use to describe connections between clusters is "weak ties." Kadushin writes that "weak ties facilitate the flow of information from otherwise-distant parts of a network."[14] Weak ties help to integrate social systems and form bridges between different clusters. This is a confusing term because of the subjectivity of the notion "weak." I prefer to consider weak ties as "connectors": individuals who are capable of bridging the gaps between segmented networks.

Social systems that lack weak ties are more likely to be fragmented and incoherent. In such cases, "new ideas will spread slowly, scientific endeavor will be handicapped, and subgroups that are separated by race, ethnicity, geography, or other characteristics will have difficulty reaching a modus vivendi." As Kadushin suggests, although there is great interest in how to create linkages between "otherwise isolated circles, groups, and organizations," social science has not yet "systematically explored the *attributes* of the nodes or vertices that produce these connections or the obverse—the social barriers that might prevent connections."[15]

This is exactly the challenge that requires our attention, and should be prominent in the minds of citizens and educators alike. Homophilic relationships are safe and familiar, and re-create the tribal associations deeply embedded in our evolutionary past. We should understand why they matter and hold such attraction. Yet we also know that innovation, new ideas, and creativity are stimulated by sharing relationships with people and networks that are different than we are. There are individuals who are particularly suited by virtue of their personalities to exercise leadership through strengthening homophilic ties or serving as powerful connectors. This is true in smaller-scale clusters or at the level of entire nations. I suggest that a better understanding of these dynamics lies at the heart of the tides of change challenge I discussed in chapter 3, the relationship between tribes and territories, or the complex dynamics of us versus them.

In a time when our access to networks, whether it's LinkedIn, Facebook, or whatever social media du jour, is most trendy, all citizens would benefit from a deeper understanding of how social networks impact their lives, and how they can take more agency in utilizing them effectively. There have been many controversies about how the analytics of Facebook (or other networks) provide political or commercial interests with the means to manipulate homophilic networks. One antidote to this challenge is to build immunity from that manipulation. And that immunity requires a commitment to civic education through a deeper understanding of social network formation and process.

Networks, Identity, and Education

Social networks play a powerful role in personal identity, helping people shape their attitudes and values within the context of peer group relationships. This is true in almost every human cultural situation. With the development of electronic social networks, the dynamics of personal identity may be shifting. Yet as Kadushin indicates, despite the increased research attention on social network analysis, the psychological foundations of human social network behavior is "surprisingly ignored."[16]

His synthesis of the available research explains that social networks serve both motivational and cognitive purposes. "One motivation is to stay within one's social cocoon, for the connections between people and social units lead to feelings of safety, comfort and support. Another motivation is to reach out and make connections where there were none." Kadushin is particularly interested in applying these motivations in organizational settings as a measure of individual effectiveness. In what situations do these motivations influence personal and organizational behavior? For example, "the person who has connections with a variety of others in different clusters also has access to information and ideas that are not present in her own cluster."[17]

The best way to think about these motivations is to reflectively consider how and whether they apply to your own social networks, and distinguish whether there is a difference in those motivations as manifested in visceral and electronic relationships. This reflective approach is especially salient because various social media outlets often reinforce like-minded perspectives, creating social and political polarization along with miscommunication and distrust between different clusters, thus impeding cultural mixing and possibly posing a threat to civic understanding. This dynamic isn't necessarily the cause of the cultural polarization rampant in contemporary culture, but it does explain why homophilic networks mainly reinforce cultural norms, lodging personal identity in a circular loop of recursive perspectives. Why listen to another point of view when it's so easy to reinforce your own predisposition? I don't think that social media is the cause of this tribally derived inclination to find safety and affiliation in homophilic groups, but it certainly can expand the membership of like-minded groups and create the illusion of safety in a world of complex moral differences.

People and organizations that understand social connectivity between segmented clusters are the hubs of innovation and creativity. That's why cities are centers of innovation. Connectivity occurs both through intention and serendipity. Steven Johnson in *The Natural History of Innovation* describes how in both organizational life and the natural world, innovation occurs in what he refers to as liquid networks. He uses the evolutionary history of life as a metaphor. "And so, when we look back to the original innovation engine on earth, we find two essential properties. First a capacity to make new connections with as many other elements as possible. And, second, a 'randomizing' environment that encourages collisions between all the elements in a system."[18]

Interestingly, those properties describe both carbon as a biogeochemical connector and cities as a social network connector. Johnson's book is filled with interesting natural history metaphors that lend insight into the convergent processes of innovation and creativity. For our purposes, there are two important points. First, the natural world can inform how we think about social connections. Second, we should make education for social connectivity a priority in organized learning environments.

How can we enhance education for connectivity? And what kind of connectivity are we talking about? I will address this in the concluding section of this chapter. As the first step, I'd like to go into more depth regarding network navigation and visualization with the following caveat. The risk of studying social networks exclusively is that we reduce our social lives to human affairs only. Many indigenous cultures would never make such a distinction. Human affairs were inextricably linked to the lives of plants, animals, and the presence of landforms and watercourses. There was no distinction between social and ecological networks. They were one and the same. Perhaps our sophistication has grown. Yet the most exciting ecological and evolutionary research suggests that the relationship between ecological and social networks is often indistinguishable. This is at the core of Alberti's urban ecology research, and why she titles her book *Cities That Think like Planets*. We enhance how we know the world when we reiterate and experience those profound connections.

Learning to Navigate Networks

Even when we cannot see the terrain clearly—in the dark of night or under the cover of snow—the road is a guide and companion and an assurance that in time we will come to our desire, whether that is adventure and new territory or the familiarity of home.
—Amy Ronnberg and Kathleen Morris, *The Book of Symbols*

Leonhard Euler, an eighteenth-century Swiss mathematician, posed the well-known (among mathematicians) "Seven Bridges of Konigsberg" problem. How can you take a walk through town, visit all seven regions of the town, and avoid crossing one of the seven bridges twice? It turns out it can't be done, and Euler proved its impossibility. In so doing, however, he laid the foundations of topology and graph theory—branches of mathematics that analyze the patterns of paths and networks.

If network thinking is becoming increasingly prominent in the Anthropocene, then we need some navigational aids for learning how to identify, respond to, travel through, and then assess the important networks in our lives. Topology can help us understand the patterns of networks, and such knowledge illuminates movement and flow. Yet there is much more to network thinking than just mathematics, interesting and beautiful as it might be. We also have to understand the paths that are available to us, how we can reconstruct them as necessary, and then how we use them in our lives.

This section is a brief guide for how to educate people about navigating networks. Let's remember that networks are structures that allow for the movement of some combination of matter, energy, or information. Ecological networks involve all these factors. That's one of the reasons they are so hard to visualize. In social networks, we mostly focus on human communication: the ideas and conversations that move between people. Learning how to navigate networks is a hands-on way to map patterns of information.

I present an eight-step process for learning how to navigate social and ecological networks. This is intended for use in any educational setting. Readers can modify the approach depending on the age, size, and background of the group, and further adjust the method depending on the subject matter under investigation. This sequence is most interesting when applied in group settings as it's instructive to share the patterns and insights that emerge. But you can also do it on your own. These are open-ended guidelines, subject to modification, improvisation, and imagination. How we teach and learn about networks is a wide-open field, and as we learn how to do so, it's important to share what we've learned.

Part I: Identification, Mapping, and Structure
Step 1: Identification
What type of network are you thinking about? Although any network structure will inevitably entail considerable intricacy and complexity, it's crucial to initially limit the scope of inquiry. You can identify social networks such as friends, professional contacts, or people who have influenced you. In organizational settings, it might be sources of revenue, clients, outreach, strategic choices and how they are connected, or an assessment of management structure. For environmental learning, you can choose food webs, biogeochemical cycles, watercourses, weather systems, root systems, or the movement of individual species through a landscape.

Step 2: Scale
Once you identify the subject, how will you define its scope? What is its size? How many links is it reasonable to identify? Are there subnetworks you wish to identify but are beyond the scope of your inquiry? Is your chosen network a subcategory of another, larger network? In every case, the network you choose will necessarily have scale limitations that will determine what you learn.

Step 3: Mapping
Create a map of the network. What's the best medium for doing so? It's often best to start with freehand drawing since the process of sketching a map enables you to internalize the relationships you trace. What does it look like? What can you learn from its structure? How can graphic approaches elucidate the most important relationships? The map doesn't have to be conventional and linear. Experiment with alternative diagrammatic or mapping forms. Develop a legend that describes the relationships on the map—use colors, shapes, loops, or icons—or whatever representations illustrate the concepts you wish to convey.

Step 4: Nodalities
In compiling the map, emphasize the connections in the networks, both within the network itself and between networks, as appropriate. Which connections are most important? What are the crucial links? How and where does the network connect to other networks? Where are you (or your organization or habitat) located within the network? How many nodes can you distinguish? Is there a way to differentiate or weight the nodes?

Step 5: Paths and Flows
Take special notice of the way things move through the network. Are there main thoroughfares and arteries? Is the flow mainly in one direction, or are there circular paths that cover most of the nodes? Is there a core and periphery? Are there loops where things get stuck? Is there flexibility? Are there multiple paths and relationships? Are there particular nodes that serve as conduits, brokers, facilitators, or perhaps obstacles?

Part II: Interpretation
Step 6: Sourcing
Sourcing emphasizes origins. Depending on your subject, is there an original center of dissemination? Are there multiple sources that feed, fuel, or

inform the system? How can you identify them? Where are these sources located in the system? For example, if you're tracking the flow of information through a social network, let's say how people get news, where are the sources of that news, and how is it modulated and influenced as it passes through the network? In the case of information, what are the biases and influences that affect the source? How is that information disseminated, transformed, and manipulated? This is an issue of increasing significance as more people get their information from fewer and fewer sources, both because of homophilic social tendencies and the potential state or corporate monopolization of news media.

Step 7: Browsing
Browsing is the process of scanning multiple networks in order to vary and then select the paths to navigate. Recall the challenge of proliferation described in chapter 4. How do you open multiple pathways of learning, scan those pathways, and then choose filters that allow you to focus on what's most important? The challenge is to find a balance between proliferation and filtering. This is true of network thinking as well. Network structures can similarly be scanned and filtered. Are there ways to scan the network so that you get information and ideas from a variety of sources? How might the user reconstruct a network so that it contains multiple sources and various paths of dissemination? Are there network structures that enhance scanning?

Step 8: Curating
Curating is the process of organizing the network structure to facilitate learning. Are there ways to organize and interpret the network to enhance the diversity of thought, clarity of focus, and reciprocity of communication? Are there network nodes, or arrangements of nodes, that are more or less structurally suitable for a curating process? To what extent is a network curated? Is it curated by virtue of its structure, the values of the sources, or a combination of both?

The depth and extent of network navigation is simultaneously fascinating and daunting. These eight steps are merely suggestions for how to teach and learn about navigation. Our understanding of these approaches will expand in the years to come, especially as network thinking becomes more prominent. Our challenge is to organize network thinking around constructive connectivity: the ability of the navigator to enhance those connections most conducive to learning. The idea of constructive connectivity is also crucial to participatory democracy—how we ensure that people have access to informed decision making—and environmental learning—how we ensure that people perceive the intricate ecological dynamics that sustain biodiversity and human flourishing. We start by teaching about network navigation. Why? Because familiarity with network navigation enhances our ability to perceive complex networks. And environmental learning requires such perceptual flexibility.

Constructive Connectivity Continued

We've had a wide-ranging discussion about network thinking and how that informs our perception of connectivity. And we've acknowledged the unfolding complexity and proliferating research that will inevitably change our understanding of the meaning as well as importance of networks. My intention is to present evocative concepts that generate insight for the sheer wonder of it, but also to promote network thinking to enhance environmental learning. I've linked the words constructive and connectivity together to attach values to what's most meaningful in this learning process.

I'll venture a definition. Constructive connectivity promotes learning that enhances an understanding of social and ecological networks, demonstrates the relationships between the two, and promotes creativity and innovation by building relationships among diverse clusters. This process requires the ability to observe networks, learning how to navigate and visualize them, and exercising discretion and choice in determining which connections promote human flourishing and ecosystem integrity. Perceiving biosphere network archetypes provides a template for understanding the evolutionary/ecological context of network formation and function. Perceiving social networks supplies an opportunity to elevate civic discourse by sharing information among different cultural groups. Understanding the interpenetration of ecological and social networks shows how as well as why cultural diversity and biodiversity are a foundation of environmental learning.

It takes courage to be a connector, a person who aspires to bring these different perspectives together, to explore how diverse cultures enrich a community and how diverse species support a thriving ecosystem. It takes courage to expose yourself to multiple perspectives, whether it's the cultural view of a person much different than you or perceiving the natural world from the perspective of a different species—thinking, let's say, like a mycelium. It takes practice and experience to exercise that courage and insight.

If network thinking is crucial to how we come to know the world, then we need criteria for how to best achieve it. What does constructive connectivity mean to you? And we need educators who allow people, especially younger learners, to explore network thinking by taking a hands-on approach to experiencing the networks in their own lives—in their neighborhoods and communities, on playgrounds, in parks, and in understanding both cultural and natural histories. We need networks of network-thinking educators who can share experiential curriculum, and who will generate new insights about how people learn and how they can cope with the increasing complexity of a networked world.

How does a deeper understanding of networks help us learn about migration along with the movement of people and species? How does it help us understand the relationship between urban and rural communities, or between the local and global? How does perceiving networks enhance our understanding of place? Please keep these questions in mind as you read the next two chapters.

6
Migration: The Movement of People and Species

Family Stories

Like many children, I came to know the world by listening to the stories of my parents and grandparents. My mother was born in 1920 on a farm on the Mediterranean island of Cyprus. Her family of Russian Jewish immigrants fled pogroms in Belarus. They were initially denied access to the United States and hence briefly settled on Cyprus. The family emigrated again when she was nine years old, moving into a cold-water flat in Brooklyn. On arrival, my mother spoke three languages but nary a word of English. She and my grandfather would tell stories of how much they enjoyed the farm, and how difficult it was to leave the farm behind, even though they knew doing so would bring more opportunities for the family.

My paternal grandmother left the pogroms in Belarus for New York City at thirteen years old, voyaging on her own, the oldest child in a family of eight. Her mother, father, brothers, and sisters all followed. Similar to many immigrants, she became involved in a fraternal organization, the Workmen's Circle, a Jewish mutual aid society that promoted assimilation to US life as well as worker rights and socialist ideals. My father grew up in a politically progressive household of working people trying to find their way in a new country. My father and mother met while working for the Henry Wallace campaign (the third-party Progressive candidate for president) in 1948.

As a youngster, I would rather go out and play ball than listen to stories of emigration and assimilation. I found rootedness by playing sports and games with other children. At the same age, my grandparents were emigrating to the United States. As a first-generation American, my father found stability through his profession and garden. My mother found stability by keeping a home. My paternal grandparents found rootedness through the Workmen's Circle, and by keeping in close touch with and proximity to their extended families. They enjoyed security, livelihood, and community in a new country, but they always told stories of where they came from and wanted to make sure I knew some of those stories too. They never took their freedom and opportunities for granted, and made sure I was aware that these opportunities could also vanish in a heartbeat.

This was the milieu of my upbringing. My identity was formed around compelling stories of Jewish migration, diaspora, and intergenerational connections. Family gatherings would often involve heated discussions of the issues of the day, always with the backdrop of their dynamic histories. Accordingly, people expressed themselves with great passion around political, cultural, and social concerns.

My family history, one that is common for immigrant families throughout the twentieth century and into the present day, depicts many aspects of global environmental change—people who move from rural to urban to suburban locations within a single generation, struggle to find ways to assimilate within new cultures, and are exposed to changing norms and expectations in short periods of time.

The issues of global migration, refugees, and immigration are prominent worldwide controversies. Climate change, environmental insecurity, and political uncertainty catalyze global migration movements, and these concerns will reverberate for many years to come. To better understand the dynamic circumstances of human migration and displacement, I read widely in a range of subjects from the global history of migration, to the unfolding and rapidly moving science of ancient DNA, to the evolution of race and ethnicity. The proliferation of literature is copious, fascinating, and challenging.

My literature survey reinforced my sense that the global migration crisis is one of the most important environmental issues of our times, and will challenge our most profound ethical and moral beliefs. Are there ways to expand our view of human migration by better understanding its context in multigenerational time, as a fundamental ecological and evolutionary component of human residency in the biosphere? Are there ways to link our personal experiences to this perspective by tracing our family histories?

In this chapter, I'll explore why I believe environmental learning is crucial for deepening our understanding of migration. First, I will share more of my family history with you by discussing and interpreting the results of additional ancestry research. That will lead to a brief look at the contemporary migration challenge and why it is such a polarizing issue. I'll explain why migration is intrinsic to biosphere processes, why globalization is characteristic of human history, and how contemporary migration issues are redefining the relationship between cities and the countryside. I'll dip into some of the latest genomic research, and how it is changing how we conceive of race and ethnicity. I'll cover several inspiring projects that seek to welcome immigrants, and suggest they are templates for schools and communities. The final section offers some mapping activities that apply many of the concepts covered in the chapter.

Geno 2.0: Do You Know Where Your Ancestors Are?

An immense land lies about us. Nations migrate within us. The past looms close, as immediate as breath, blood, and scars on a wrist. It, too, lies hidden, obscured, shattered. What I can know of ancestors' lives or of this land can't be retrieved like old postcards stored in a desk drawer.
—Lauret E. Savoy, *Trace*

Every week, or so it seemed, while in the process of writing this book, I would receive another email from *National Geographic* with a special offer on Geno 2.0, its proprietary ancestral heritage package. At the same time, the issues of global migration, refugees, and all the surrounding controversies were becoming heated as well as polarizing.

I began taking these Geno 2.0 advertisements more seriously. I sent away for the package, received it, spit some saliva into a small vial, sent it back, and awaited the results. During the waiting period, I attended the Money in Place conference in Taos, New Mexico—a project designed to catalyze local investments, grounded in community entrepreneurship. One of the participants at the small gathering was a Jewish woman from Boston. She was particularly interested in diversity and social justice. She told me she'd recently received her Geno 2.0 results. She was hoping it would reveal an exotic mixture, thus providing evidence of racial diversity in her own background. She looked at me and asked if I would guess the results. I told her I had no idea. She said she was 99 percent Jewish diaspora. I found it inspiring that during a time of the dangerous resurgence of white nationalism and its accompanying myths of racial purity, a person who was such an avid supporter of diversity was hoping she had a more diverse ancestry. She was disappointed she had a seemingly straightforward heritage.

My "Regional Ancestry" (550 to 10,000 years ago) results came shortly thereafter. I was 83 percent Jewish diaspora and 17 percent southwestern Europe. I assume that the southwestern Europe percentage is partly from Sephardic Jews and the Jewish diaspora is mainly Ashkenazi, but there is no way of reliably assessing the accuracy of these results, as they are still highly generic and only provide the broadest brushstrokes of your hypothesized ancestry.

The Geno 2.0 results also include your "Deep Ancestry" (1,000 to 100,000 years ago) of your maternal and paternal lineages, dating back to your African origins. On my mother's side, the path is reasonably straightforward. Many generations resided in West Asia (for approximately 60,000 years) before they departed for Europe about 8,000 years ago. On my father's side, the lineage includes about 15,000 years in Southwest Asia, followed by a move to South Asia about 45,000 years ago, Central Asia 30,000 years ago, West Asia about 15,000 years ago, and then finally Europe. These results confirmed my intuition, as my mother had strong European features, but her father and brothers had Mediterranean features. My father and grandmother had features that hinted at Asian origins.

The results are broad and general. They elucidate genomic patterns rather than the details of specific genomic lineage. If we assume that over that 100,000-year span, the average generation is twenty years, then the entire story encompasses 5,000 generations. Over the 1,000-year span, we can estimate 50 generations. Within the range of the conceivable, I have access to at most 5 generations of Thomashows, Wolfsons, Ginsbergs, and Naginskis, from my granddaughter to my great-grandmother Wolfson. I don't know family names from earlier generations. Even the fifth generation is beyond my grasp and experience. Fifty generations seems unattainable. It's not difficult to remember fifty things. Still, it would take an extraordinary effort of research to reach back even 10 generations. Thinking about 5,000 generations is virtually inconceivable. Families that can trace their ancestry further back in time either have the resources and time to do the necessary research, or have pretty much remained in the same place. They are in the distinct minority. In a country colonized by the labor of migrants, immigrants, and slaves (yes, that would be the United States of America), it's no surprise that genealogical records are sparse. Yet the science of DNA provides an extraordinary new and emerging portfolio of data that will be strengthened as its database expands.

What does any of this mean? I haven't learned much new about myself. But I did gain a greater appreciation for the complexity of human migration. My understanding is that after you've gone back six generations or so, chances are you are a mixture of multiple ethnicities, and your DNA reflects that. As Adam Rutherford suggests, "The truth is that our pedigrees fold in on themselves, the branches loop back and become nets, and all of us who have ever lived have done so enmeshed in a web of ancestry."[1] The overwhelming conclusion is that notions of race and ethnicity are cultural constructions—ways to distinguish ourselves based on short-term histories of territorial advantage, perceived wrongdoing, and a generalized fear of outsiders. This isn't to deny that cultural and genetic heritages may have distinguishing characteristics, but they are constantly changing, forever impacted by the swirling movements of people on the ever-dynamic landscape of the earth.

I take pride in my Ashkenazi ancestry because it helps me understand who I am and who came before me. I respect the best wisdom that emerges from that legacy, although I am sure there are ancestors and relatives for whom I would be greatly embarrassed. I take equal pride in the physical and cultural environments that have shaped me, and I do so out of an abiding respect for the ecological and evolutionary inheritance that shapes all aspects of my being. Our genetic ancestry in relationship to the ecological/evolutionary matrix is a magnificent epic narrative that should unify all humans, as it is the ultimate story of interconnectedness and what Buddhist philosopher Thich Nhat Hanh describes as interbeing.[2]

But alas we are living in an era of resurgent nationalism, despite this groundbreaking DNA research and our deeper understanding of human interpenetration. The insider/outsider dynamic is essential to our heritage as well. It's our responsibility to better understand the danger and potential of cultural pride. I'll suggest, as we proceed through this chapter, that a better understanding of our deep human ancestry contributes to a moral ground for welcoming migrants and refugees, and is a crucial foundation for environmental learning. Before we proceed with the conceptual sequence regarding migration and environmental learning, let's look more closely at what terms such as migrants, refugees, and immigrants mean.

Refugees, Migrants, and Immigrants

In the first two decades of the twenty-first century, the refugee/immigrant/migration challenge spawned weekly crises of conscience. Consider the long list of catalysts: people who lost everything after extraordinary weather catastrophes, earthquakes, or tsunamis; people fleeing civil wars or brutal regimes, people escaping from gang violence and drug wars; or people who were economically desperate. Difficult political controversies ensued, prompting the full range of human emotional responses, from empathy and compassion, to nationalism and anti-immigration sentiments.

In June and July 2018, sparked by a resurgence of anti-immigration policies in the United States, a full-fledged border crisis emerged, culminating with the heartless separation of families at the Mexican border. Here's a snapshot of headlines in the *New York Times* during the week of July 2, 2018:

"'Tracking a Package': The Perils, and Price, of Migrant Smuggling"
"Affluence of Bavaria Is Belied by the Anger"
"Spain's Migrant Wave Grows, Tripling in 2017, Even as Europe's Subsides"
"This Italian Town Once Welcomed Migrants. Now It's a Symbol for Right-Wing Politics"
"Why Europe Could Melt Down over a Simple Question of Borders"
"'Fraternite' Brings Immunity for Migrant Advocate in France"
"Scorpions, Dehydration, Disease: Syrians at the Border Face Deadly Threats"
"When a Baby Is an Everyday Reminder of Rohingya Horror"
"In a Migrant Shelter Classroom, 'It's Always like the First Day of School'"
"Trump Administration Says It Needs More Time to Reunite Migrant Families"

Refugees, migrants, and immigrants. Who are these people? Where do they come from? How can they be treated with respect and dignity? How can they find new homes and opportunities? How can we better understand that for a simple twist of fate, they could be us? Or that we may be or have been only a generation or two removed from suffering similar fates (in the past or future).

These headlines remind us of the ubiquity of these questions. They are not unique to our time. Migration and displacement are intrinsic to the human condition. But in the Anthropocene, as a consequence of economic globalization and rapid transportation, we are all so much closer together, and the plight of global refugees potentially impacts every community. The idea of borders, like race and ethnicity, is merely a human construction that concocts the illusion of safety, a territorial expression, and the demarcation of boundaries to distinguish between insiders and outsiders. No wall or boundary, though, can possibly be big or strong enough to hold back the tides of change in a world of perennial movement.

Before we proceed any further, let's parse these terms. The UN Refugee Agency distinguishes between refugees, internally displaced persons, stateless persons, and asylum seekers. A "refugee" is "someone who has been forced to flee his or her country because of persecution war, or violence. . . . Two-thirds of all refugees worldwide come from just five countries: Syria, Afghanistan, South Sudan, Myanmar, and Somalia." Over half are school-age children under the age of eighteen. An "internally displaced person" is "someone who has been forced to flee their home but never cross an international border." The United Nations estimates there are 40 million internally displaced persons in the world. A "stateless person" is "someone who is not a citizen of any country." The United Nations estimates 10 million "around the world are stateless or at risk of statelessness." "Asylum seekers" are "people who flee their own country and seek sanctuary in another country." There were 1.7 million new asylum claims in 2017.[3]

Any of these people may be or will at some time become migrants, defined by Human Rights Watch as "people living and working outside their country of origin." As this nongovernmental organization points out, "Large numbers of migrants fleeing criminality, poverty and environmental disaster will be without the protections of refugee status."[4] According to the Pew Research Center (as of 2016), "If all of the international migrants . . . lived in a single country, it would be the world's fifth largest, with around 244 million people.[5] In 2017, according to the United Nations' *International Migration Report*, there were 258 million migrants.[6] The International Organization for Migration, another UN activity institution, has a website that lists thirty-four pertinent migration terms.[7] Migration, it seems, is complicated enough to generate a unique suite of designations.

An immigrant is a person who resettles in another country. This entails an immigration process that varies in its legal protocols from one country to another. Some immigrants come by virtue of choice, pursuing job opportunities, and they may be recruited by employers. But migrants are almost always people who are displaced. When we discuss immigration policy, all these issues matter, and the distinctions are crucial.

Inevitably the plight of refugees and migrants places great pressure on potential destination hosts. Those who are opposed to immigration are concerned that the influx of newcomers will diminish economic opportunities for people who reside in the destination country, will require too many of the destination country's resources to accommodate them, or will bring unsavory elements, including crime and/or terrorism. Some of this is prompted by a generalized fear of outsiders, and some comes from a legitimate concern regarding their own economic well-being or perceived threats to cultural integrity. Migration waves often spawn a resurgence of nationalism and/or racism, an ensuing call for stricter border controls, and/or outright discrimination regarding migrant populations that already have taken up residence in the host country. These are the underlying dynamics of all the newspaper articles cited above.

In chapter 3, I discussed the intricate convergence, or interconnected dynamics, of environment, equity, diversity, and inclusion. I briefly explained how a significant percentage of migrant populations are displaced because of environmental catastrophes, frequently exacerbated by political decisions. Hence the political, cultural, and social outcomes of environmentally induced migration must be considered. The migration/refugee crisis accompanies an environmental crisis. And as the issues become more serious, as will likely be the case with changing climate and sea level rise, the racist, exclusion-oriented approach will continue to cast its dark shadow. There is no escaping the urgency of this dilemma.

Why Environmental Learning about Migration Matters

The key is to provide or somehow create among people stronger clues of trust and common values than might otherwise be suggested by the highly imprecise markers of ethnicity or cultural differences that we have used throughout our history, and then to encourage the conditions that give people a sense of shared purpose and shared outcomes. That is the recipe that carried us around the world beginning around 60,000 years ago, and it still works. Looking around the great cosmopolitan cities of the world, it is hard to avoid the conclusion that this is already happening.
—Mark Pagel, *Wired for Culture*

Pagel's optimistic message comes at the conclusion of a comprehensive discussion about the evolutionary origins and cultural manifestations of the human social mind. I find Pagel's work of great interest because he explores the complexity of social groups along with the precarious relationship between four kinds of social behavior: altruism, selfishness, spite, and cooperation. The origins of these behaviors is a fascinating, unfolding, and dynamic research agenda, easily co-opted by ideological predispositions. To understand how we respond to migration (in all its forms), it's crucial to reflect on the full gamut of social responses. Such reflection is a prerequisite for global citizenship in the Anthropocene because of the inevitable proximity of cultures and flow of people that makes it happen. The political conflicts may play themselves out in the dangerous emotions of us versus them polarizations. How do we avoid this?

In a previous section, "Tribes and Territories," I suggested that it may be impossible to overcome the evolutionary and cultural legacies of us versus them dynamics, but it may be possible to mitigate them. As this is a book about environmental learning, the challenge is how to educate about these issues, expand awareness, situate migration as an ecological response to living in the biosphere, better understand the ecological and cultural dynamics of migration, and cultivate empathy for the movement of people and species. I conceive this in six steps: from the biosphere to social behavior—from a map of the world to your backyard or city street, and then to your personal experience. Somewhere in your past, present, or future, your family has a migration history.

1. Migration is a biospheric phenomenon. Continents move. Atmospheric and oceanic circulations are dynamic flows of air and water. Biogeochemical cycles move nutrients through air, water, and land. Lifeforms navigate these dynamic flows by finding habitats and food sources, while constructing networks and forming behaviors that lead via evolution to an infinite variety of species and adaptations, many of which involve moving from one place to another.

2. Migration is intrinsic to the human condition. From the dawn of humanity, roving bands moved from place to place in search of food, security, and habitat. Long before there were borders on published maps, these bands established territorial boundaries in a matrix of coexistence and conflict, necessarily prompted by their inevitable movement. As the environment constantly changed, survival demanded multiple strategic approaches, and one of the most important was the ability to follow the food sources.

3. Recorded human history is a convergence of multiple ecological and cultural migrations. People disperse in response to environmental change and dynamic territorial boundaries, forming distinct tribal affiliations that lead to diverse cultural worldviews, power imbalances, and the possibilities of exclusion, inclusion, assimilation, displacement, or retreat.

Flip through the pages of a good historical atlas to observe the constant rearrangement of borders, empires, ethnicities, and civilizations. Notice especially the waves of invasion and imperialism as well as the expansion of trade and cooperation.

4. The Human Genome Project is ripe for interpretation. With the proliferation of genomic data, we are gaining a much better understanding of both human history on the planet, especially regarding migration and movement, and our personal and ethnic ancestry. It's becoming increasingly clear that race and ethnicity are cultural constructions. Go back in time just a few generations and you can review the wonderful diversity of your ancestry while also tracing the geographic path of your genome.

5. Empathy, compassion, and altruism emerge from expanding personal identity into the collective sphere, and this includes overcoming tribal affiliations and the prejudices they promote. The Anthropocene is characterized by both the intense impact of human activity and the increasing proximity of cultures and ecosystems. There is no escaping from the inevitable political challenges of such proximity. These challenges easily provoke us versus them dichotomies, nationalist impulses, and the building of walls and obstacles. How do we maintain cultural integrity, political sovereignty, and a pride in local place while enhancing empathy, intercultural understanding, ecological resilience, and constructive connectivity?

6. The most successful, innovative, and enduring cultures understand the necessity of cooperation, trade, and exchange.[8] Cultures and ecosystems alike thrive when they are most diverse. This requires permeable borders, a spirit of exchange, and an understanding of the conditions that contribute to cultural and ecological diversity.[9]

Migration and the Biosphere

Every hour of every day, somewhere, some place, animals are on the move—flying, walking, crawling, swimming, or slithering from one destination to another. It is the ancient ritual of migration, and it is happening everywhere.
—David Wilcove, *No Way Home*

Wilcove, an evolutionary ecologist, explains that "tens of thousands of species migrate, and the journeys they take are as different as the creatures themselves."[10] His excellent book explores the ecology of animal migration, covering an extraordinary array of critters—the circumpolar voyages of Arctic terns, the journeys of Costa Rican three-wattled bellbirds from montane cloud forests to lowland jungles, the great white sharks that wander halfway across the Pacific Ocean, and then all manner of birds, butterflies, dragonflies, mammals, and fish.

To get a firsthand glimpse of these magnificent species' journeys, take a look at the book by James Cheshire and Oliver Uberti, *Where the Animals Go*. Using the most sophisticated tracer and mapping technologies, they create a splendid portrait of fifty species' migration paths, from warblers who dodge tornadoes to the remarkable journeys of sea turtles. Their final chapter discusses how we might use the same technology to trace the patterns of human migration. Right now, as you read this, dozens of migratory species are moving from one place to another in search of food and habitat, using a variety of senses to assist them in their travels. Check out Cheshire and Uberti's maps to expand how you think about migration. A Google Image search for wildlife migration maps is similarly illustrative.

And as evolutionary biologist Geerat J. Vermeij reminds us, globalization, by which he means the migratory dispersion of species, is "by no means unknown in the pre-human biosphere." At various times during the last 50 million years, global migration was facilitated by the movement of continents along with shifting oceanic and atmospheric circulations, or what he describes as "previously unreachable land or ocean barriers." The most spectacular of these connective pathways were the Beringian land bridges (which alternately opened and closed starting 50 million years ago), allowing mammals and ultimately humans to pass from Eurasia to the Americas, and then approximately 3.5 million years ago, the "emergence of a continuous isthmus in Central America," enabling the spread of plants and animals between previously isolated continents.[11]

Vermeij is mainly interested in how this global migration dynamic is an adaptive response to changing environmental circumstances, most specifically the search for food, with the implication that the destination has more productive supplies than the original habitat. Migration became a way to move between seasonal production cycles. Keep in mind these land bridges dramatically changed the ecological balance of the destination. As Vermeij suggests, "With the establishment of a continuous Central American corridor, North American invaders changed the island continent of South America into an extension of the North American land mass, and through conquest raised the competitive bar there."[12]

There's no reason to restrict yourself to the last 50 million years. Browse through Ron Redfern's incredible book *Origins: The Evolution of Continents, Oceans, and Life* to experience a dazzling display of perennial geologic and biological movement, infinite patterns of oceanic circulations, atmospheric changes, landscape configurations, ecological relationships, and evolutionary adaptation. Pick any era of earth history, and you'll find continents and species on the move. Wait around for 10,000 years (or less) and biogeographic relationships will undergo significant change. The biosphere is an ever-changing matrix of dynamic movement at every conceivable temporal and spatial scale. I'll have more to say about this in the forthcoming chapter on improvisation and environmental learning.

Both Wilcove and Vermeij are interested in how anthropogenic change will impact migration dynamics. Wilcove writes that migratory species are in decline as the corridors of movement are increasingly threatened. Vermeij discusses how the rise in ocean temperatures alters the productivity and movement of phytoplankton, changing the biogeochemistry of oceans and hence the availability of food supplies. They present more evidence that the Anthropocene is an era of unprecedented environmental change and speculate about the requisite adaptive challenges. It's not the change itself that's unprecedented; it's the rapidity with which it is occurring. Scientific fields such as conservation biology, environmental change science, and resilience studies (among many others) are striving to understand what policy interventions are necessary, both to preserve species habitats and allow for human flourishing.

Placing human migration in the context of the biosphere surely broadens our perspective, and at the very least demonstrates the ubiquity of migration as an adaptive strategy. But it is a wondrous process too. As Wilcove observes, "Almost every aspect of migration inspires awe: the incredible journeys migratory animals undertake and the hardships they face along the way; the complex mechanisms they use to navigate across the land and through the skies and seas; the sophisticated tools with which scientists study them; and not least, the perseverance of the people striving to save these animals in the face of an increasingly congested, inhospitable world."[13] It's about time we consider human migration with the same sense of awe, wonder, care, and significance.

The Original Human Globalization

They have been walking from the beginning,
through the foggy sponges of lowland

forests, under umbrella leaves, in the shattered
rain of ocean beaches, through the tinder
of ash pits, the thickets of cities, along washes
and ravines and the dust of dry creek beds.
—Pattiann Rogers, *Generations*

In *Who We Are and How We Got Here*, David Reich describes how the "new science of the human past" inspired by the genome revolution provides more precision regarding the human journey, from approximately 160,000 years ago to the present. In the years to come as DNA sequencing and the fossil record supply more data, we'll increase our understanding of this compelling narrative. Reich suggests we are in the early days of a profound revitalization of how we understand the evolution of humanity. He foresees "an ancient DNA atlas of humanity, sampled densely through time and space. This will be a resource that I think will rival the first maps of the globe made between the fifteenth and nineteenth centuries in terms of its contribution to human knowledge."[14]

Current research and theory indicates the human species lived precariously some 160,000 to 200,000 years ago, *Homo sapiens* competed or coexisted (it's not entirely clear) with a multiplicity of human lineages, including most prominently Denisovans and Neanderthals over the next 100,000 years, and *Homo sapiens* experienced an evolutionary breakthrough between 120,000 and 60,000 years ago, which eventually led to a series of migrations out of Africa. One of the exciting and controversial areas of this research involves the various hypotheses as to why this occurred, with explanations that include various combinations of symbolic thinking, social learning, the emergence of feelings, and the origins of language.

Once this breakthrough took place, humans spread throughout the entire globe in an amazingly short time, especially considering they mainly did this by walking. Pagel, in *Wired for Culture*, traces the rough outlines of what was a series of migrations, culminating in a global diaspora that was by and large complete, with the exception of Pacific Ocean settlement, by approximately 15,000 years ago.

The modern occupation of the world was now complete, all within a few tens of thousands of years after leaving Africa, and most of this within the first 20,000 years. It was an occupation that had begun back in Africa when as few as several hundred to several thousand people left that ancestral continent, so that today, remarkably, and in such a short period of time, all of us on earth trace our ancestry to this small and intrepid band.[15]

This included the inhabitation of every conceivable environment—from deserts to arctic to jungles—with thousands of different ecological survival strategies, and the consequent differentiation of languages, mating practices, types of shelter, symbolic communication systems, and worldviews. As Pagel writes, "A single species had acquired a global reach, and specialization, lifestyles, and beliefs as varied as collections of different species."[16]

Just pause your everyday activities for a moment, and suspend your inevitable assumptions about whatever predominant worldview you ascribe to, the language you speak, and the currency you use, and consider that, according to Pagel, "there are currently as many as 7,000 different languages spoken, or 7,000 mutually unintelligible systems of communication in one species, marking out at least 7,000 distinct societies. This is more different systems of communication in a single mammal species—for that is what we are—than there are mammal species."[17] For a good visual summary of this process, refer to the *National Geographic*'s Genographic Project's "Map of Human Migration."[18]

To remind yourself of the vastness of these incredible migratory journeys and great differentiations that followed, it's helpful to regularly scan the pages of a good contemporary atlas (I like the tenth edition of the *National Geographic Atlas of the World*). Every settled place, every town and city, is an artifact of this process. When we create borders between nations or develop rationales for limiting some movements, we stem the tide of the great legacy of human migration, which is a response to the dynamic movements of the biosphere. When viewed at expanded spatial and temporal scales, human migration is yet another grand, global circulation. This great flow continues unabated in the Anthropocene, and it will increase with the severity of environmental dislocations.

It's equally instructive to trace the flow of contemporary human migratory flows. The most robust flows occur from poor nations to wealthy ones, from places of political turmoil to those of political stability, and from environmental catastrophes to any place that is temporarily stable. This creates great pressure on the affluent nations. Indeed, the one-way migratory flow creates instability in the destination countries as these places are often polarized about how and whether to accommodate the newcomers.[19]

Converging Migrations and the World City

Transnational cities are prime destinations for international migrants and they are highly diverse. The mobilizing force of the transnational city is primarily expressed in its regional appeal: it is a node on the global landscape, highly visible, attractive for various reasons, and drawing hundreds of thousands of migrants.
—Paul Knox, *Atlas of Cities*

The world's population shifts cityward in a back-and-forth oscillation of single individuals and clusters of villages, pushed and pulled by tides of agriculture and economy, climate and politics.
—Doug Saunders, *Arrival City*

The subtitle of Saunders's book captures the essence of this section: *How the Largest Migration in History Is Reshaping Our World*. Saunders explains the dynamics of what he describes as "arrival cities," places where migrants journey to expand their economic opportunities, escape political oppression, and provide more options for mobility and education. He profiles over thirty arrival cities, telling the stories of the people who travel and eventually settle there. Five important patterns (among many) stand out.

First, migration is a two-way flow between the world city and village. It is a reciprocal, dialectic movement, which urbanizes the village as much as it revitalizes the city. It serves as a sorting and selection mechanism, leaving the most ambitious and able in the city, with a large number—typically about half of all rural-to-urban migrants throughout history—returning to the village for good.[20] Second, arrival cities are often centers of innovation and entrepreneurship. Spurred by aspiration and connectivity, the most successful arrival cities are a "key instrument in creating a new middle class." They serve multiple functions, including the creation and maintenance of networks, provision of protection and security, enhanced prospects for property ownership, small businesses, and higher education—or what Saunders calls the "urban establishment platform"—and social mobility path.[21] Arrival cities serve as nodes for social capital formation. Third, arrival cities can facilitate the success of migrant prosperity by developing public policies that stimulate this social capital development. Fourth, there is a cohesive and dynamic interconnectedness between arrival cities and rural villages, and migration policy should focus on these connections. And fifth, migration patterns might be regional (within the borders of a nation) or transnational, and we need policies for both pathways.

Indeed as Knox points out, some cities are magnets for migrants, or what he calls "transnational cities." In the United States, Miami "receives more foreign immigrants than any other city" as over a five-year period it typically receives about 285,000 foreign immigrants, and then another 270,000 immigrants from elsewhere in the United States.[22] Also, approximately 315,000 people depart Miami for other US destinations. An amazing 43 percent of Miami's metropolitan population is foreign born.

Statistics from 2013 (remember these numbers are volatile), rank Miami second to Dubai, where an astounding 83 percent of the population is foreign born. The six other transnational cities of note are Vancouver, Hong Kong, Los Angeles, Amsterdam, Sydney, and Singapore, all highly prosperous, vital, and dynamic cities.

The Globalization and World Cities Research Network, an international collaboration of urban scholars, publishes a comprehensive and methodologically rigorous classification of cities, ranking them according to the complexity and depth of their international economic, social, and financial networks. The latest ranking (from 2018) includes four categories of alpha cities, three categories of beta, and three categories of gamma, differentiated according to their level of global integration.

London and New York show up as alpha++ cities; Beijing, Dubai, Hong Kong, Paris, Shanghai, and Singapore are alpha+; and Chicago, Los Angeles, Miami, and Toronto are the North American representatives among twenty-five global alpha cities. I highly recommend that readers review these lists as they provide a fascinating perspective on the depth of urban interconnectedness.[23]

A reminder that according to the United Nations Department of Economic and Social Affairs, around the time of this book's publication (2020), approximately 55 percent of the world's population lives in cities, and by 2050, this percentage is expected to increase to 68 percent.[24] Clearly a significant percentage of this increase will consist of various migrant populations. That's why there is a proliferation of research in urban studies, global cities, urban ecology, and sustainable development, and why that research must be prominent in how we think about environmental learning. The Anthopocene reflects the urbanization of the world, and that urbanization consists of the convergent human migrations of people from every conceivable place on the planet.

It should also be apparent that the division between urban and rural, and all the perceived cultural, political, and economic polarizations that result from this, must be interpreted at a larger scale as the inevitable movement of peoples around the planet. We must consider the intricate relationship between global cities and rural villages as a dynamic flow of natural resources and social capital. It's too easy to generate awkward stereotypes regarding urban and rural attitudes. Saunders claim that the "most ambitious and able" remain in the city reflects an urban bias. Rather, we must find ways to demonstrate the crucial relationship between cities and the countryside, how they both inform each other, and how we can ensure the most dynamic two-way flow of social capital.

Human migration exists simultaneously with the movement of other species along with their dislocations due to urban growth, environmental change, and natural resource extraction. That's why it's essential to perceive human migration as a biosphere process. Understanding migration is a crucial foundation for urban ecology, and prerequisite for an understanding of urban and rural sustainability. Next I'll cover how genomics potentially broadens how we think about migration, ultimately challenging notions about race and ethnicity, and hopefully dissolving some unfortunate tribal dichotomies.

The Genomic Legacy

We now know that nearly every group living today is the product of repeated population mixtures that have occurred over thousands and tens of thousands of years. Mixing is in human nature, and no one population is—or could be—"pure."
—David Reich, *Who We Are and How We Got Here*

We should ask ourselves *why* nearly all the people playing on basketball courts are said to be one of the same two colors as piano keys. For one thing, no human being has a complexion that is fully black or completely white. And all these players, whether white or black, have a light and dark side of their hands. In addition, bifurcating these or any people subtly negates the underlying unity of humankind and its common genetic and historical roots.
—Winthrop D. Jordan, "Historical Origins of the One-Drop Racial Rule in the United States"

In his exceptional essay, published posthumously and edited by David Spickard, Jordan describes the "one-drop racial rule," a designation particular to the United States that essentially labels anyone with African American ancestry, no matter the complexity of their heritage, as black. Jordan explains the sociological, demographic, and historical reasons why this is not the case in Central America, South American, and the

Caribbean, or as he suggests, "People living in these areas manage to exist and even thrive in a teeming spectrum of phenotypes and physiognomies."[25] Instead, extremely complicated social stratification systems emerged, based on multiple designations. He cites an example of a list developed by the viceroyalty of New Spain, or what we now call Mexico, that lays out sixteen (!) possible designations depending on the perceived ethnicity of the parents—Spaniards, Indians, Mestizos, Moriscos, Lobos, Coyotes, and so on. Jordan observes that this is a vestige of the feudal thinking of the Spanish bureaucracy. Still, white (and European) superiority reigns supreme: "These Latin-dominated cultures share with the United States a common presumption: the lighter the better and the darker the worse. They reflect a similar assumption of superiority for people of European background, but they do not share with the United States a rigidly bifurcated system of classifying human beings."[26]

Please compare this with Reich's comment above regarding the inevitable mixing of human populations. Or as Rutherford notes in *A Brief History of Everyone Who Ever Lived*, "Genetics has revealed that human variation and its distribution across the planet is more complex and demands more sophisticated squinting than any attempts to align it with crude and ill-defined terms like race, or even black, or white." Rutherford concludes that "from the point of view of a geneticist, race does not exist."[27]

Why, then, in the face of this science, does race still prevail as such a painful and polarizing issue? Surely our perspectives are circumscribed by the inevitability of us versus them dynamics, and their tribal or territorial manifestations—impulses that are easily manipulated by demagogues and ethnic nationalists. That's why I strongly advocate for access to genomic heritage data, and why their interpretation should be an educational imperative. Perhaps as more people are enlightened about their complex origins and recognize the depth of human intermixing, they will be less concerned about separation based on the delusion of race. At the very least, it can open public discussions and dialogues.

As the science of genomics continues to synthesize the expanding array of data, we will be able to pinpoint human origins with increasing precision. I believe it's inevitable that we'll be able to link human migration, genomic-based ancestry, and environmental change, including the coevolution of human migrants and other migratory species. Possibly too, this conceptual integration will be enhanced with a deeper understanding of human microbiota.[28] The human ecosystem is both within and without, and it can't be separated from the dynamic fluidity of environmental change in the biosphere.

My deepest hope is that education about migration, the human genome, and all the accompanying variation and diversity inspires curiosity, wonder, and respect, rather than stratification and distrust. Ideally, these positive qualities should cultivate empathy and compassion.

Empathy, Compassion, and Education

Public awareness of migration increases not only through the drama of border activities but also the settlement process. At every phase of resettlement, migrant families, especially those that are experiencing displacement, spend time in a new place, either a temporary stop or a more permanent community that may be new to them but is someone else's current home place. This adds understandable tensions, depending on the stability, prosperity, and quality of life in the community. As you would expect, host communities exhibit the full range of social and emotional responses—from prejudice and fear to welcoming accommodation.

Schools are often on the front lines of these encounters. The challenge of education for migrants and refugees is receiving increasing attention, including numerous reports and projects. The UN Refugee Agency's report *Left Behind: Refugee Education in Crisis* describes the experiences of over six million global refugee children between the ages of five and seventeen, and the ongoing efforts to provide educational opportunities.[29] An organization called Welcoming America published a superb curriculum guide, *Building Welcoming Schools: A Guide for K-12 Educators and After-School Providers*, filled with outstanding activities for building shared narratives.[30]

Possibly the most effective way to cultivate empathy is to better understand the lived experiences of migrant and refugee children. There's an emerging catalog of children's books that highlight the extraordinary journeys of these children, emphasizing their struggles, hardships, and heroism. Just a brief literature search yields dozens and dozens of children's books addressing these topics.[31]

This is an inspiring development, and reflects well on some of the groundbreaking educational efforts that help us better understand the human dimensions of this crisis. All these books cultivate empathy, compassion, and concern, allowing readers to relate to the complexity of the challenge through the lens of personal identity. It was that same compassion that prompted several hundred thousand demonstrators (in July

2018) to protest the Trump administration's border policy. Education and political awareness reinforce the necessity of public outcry as well as action.

Inevitably, nations must develop coherent, morally sound, historically contextual, and ecologically based policy for facilitating the movement of peoples between places, especially to accommodate the displaced and marginalized. At that scale, policy debates are frequently manipulated for short-term political advantage, and it's hard for people to fully understand the complexities of the challenge. At the individual and community scale, at a place-based level, it's more likely that people can come together and find a common ground for dealing with newcomers. This won't always be so, and there are examples of communities that would rather not accept outsiders. Still, there are just as many instances of communities that find ways to welcome the displaced, and exercise kindness and empathy for people who are forced to leave home and have nowhere else to go. These are the narratives that we should emulate, amplify, and explore, knowing that the success stories appeal to the human heart and open the doors of compassion.

There are curricular approaches for doing this. In the next section, I suggest a brief catalog of mapping and interview activities that aspire to expand awareness of the movement of people and species.

Mapping Migration: From Biosphere to Psyche

Three conceptual processes are necessary to expand personal (and then collective) awareness about migration—understanding scale, exploring community, and cultivating empathy. There are many instructive ways to approach this, and much depends on the group you are working with. We have a growing body of migration-related curriculum, some of it designed to work with refugee populations or recent immigrants to facilitate the assimilation experience, and some to build community awareness among nonmigrant populations. This is an important emerging field in both public and community education, and we need to support research and practice accordingly. Since this is a book about environmental learning, however, that is the perspective I'll be taking here. Similar to all the curricular suggestions in this book, I'm providing a template for learning, and these ideas should be adapted depending on the relevant learning circumstances. These activities are organized around place, mapping, narrative, and memory. Each of these templates can form the basis of extensive courses and projects. You can pursue them in any sequence. Or if you are not an educator, consider them as wonderful learning investigations to supplement this chapter.

1. Migratory Pathways in the Biosphere: Understanding Scale

Start by mapping all the movement in your place, wherever you might be, going as far back in time as your imagination will allow. It's instructive to provide broad parameters; consider everything that moves through your place, and then see what emerges. I also find it helpful to trace patterns in multiples of ten years back in time (ten, a hundred, a thousand, ten thousand, and a hundred thousand years ago). Here are some examples.

You can trace the flow of weather patterns. Where does today's weather come from? Where does today's air originate? Where was it yesterday, and where will it be tomorrow? Can you visualize the weather systems that create today's weather?

You can trace the changing patterns of landscape. What did this place look like at various temporal intervals? How did the climate, geomorphology, and biogeography change?

You can trace the patterns of species migration that accompany the changing landscape. Are there seasonal migration patterns—salmon, birds, butterflies, or bats? What species lived here at specified time intervals? What was their range then, and what is it now? What species have gone extinct? Which are considered invasive species, and how did they arrive here?

You can trace the patterns of human migration. Who lived here? When were they here? How long did they stay? How did they interact with the landscape? How did they coexist with other species? What other peoples did they interact with?

These activities, taken together, broaden the scale of migration, so it can be seen as more than a contemporary phenomenon. They raise many political, ethical, and even spiritual questions about the meaning of origins, who makes claim to a place, and the shared human and ecological heritage of a region. Most important, they serve to place migration in both an ecological and evolutionary context.

An excellent resource for this inquiry is the book by Eric W. Sanderson, *Mannahatta*, which offers multiple landscape portraits of the ecology of New York City through time.[32]

2. Migration in Your Human Community: Exploring Migration Narratives

Depending on where you live, decide on three reasonable scales of inquiry. In a dynamic urban neighborhood, it might start with a city block, expand to include ten-square blocks, and then encompass the entire city and metropolitan area. In a rural area, distances for appropriate inquiry will be much different. Here I'll focus on an urban perspective. Similar to the migration and the biosphere activities above (mainly organized around time), construct spatial configurations that enable you to understand how your home place connects to a larger environment. At each scale, create a mapping portrait of the various ethnicities. Then juxtapose that portrait with species movements. Seek out the personal (and species) stories that are most interesting. They compose a tapestry of ecological and community narratives.

At the city block scale, create a portrait of the various neighborhood ethnicities. The challenge is how to get the information. Some of it might be available through census data, which can be helpful, but it's also interesting and vital to conduct a series of interviews with people in the neighborhood. It may be impossible to get a comprehensive picture, but you can make some choices about who to engage and why. Such choices can be made based on age, proximity (people who live in your building), and gathering places (community centers and shops), or by speaking with people you already know and asking them for suggestions about who to speak with. You might ask questions such as, How long have you lived here? Where did you live previously? Tell me about every place you've lived since you were born. Where are your parents from? Your grandparents? Are there migration experiences within the last two generations of your family? There are many ways to represent this information, but I think it's most useful to compile oral interviews through artistic mapping, photography, and film, or whatever milieu seems most appropriate.

At the ten-square-blocks scale, you can create a similar portrait. Obviously different groupings will be possible, as within ten blocks there may be a variety of communities, each featuring different ethnicities. Similar to the city block scale, you can rely on selected interviews with several people in distinct communities, or trace the patterns of movement among and between different ethnic groups. Much of this information will come through personal interaction, as the main point is to gather the migration stories of as many different people as possible. Another approach, especially in an educational setting, is to designate different areas of a city for different researchers and then develop a composite map. Or different researchers can focus on the same area and compare their experiences.

Finally, at the citywide scale, it's interesting to create a composite map of the amazing varieties of people. A temporal perspective will illuminate the changing composition of the city, how neighborhoods change, who lives there over time, where people come from, and where they go when they leave. Given the dynamic urban gentrification process, this is a revealing way to trace patterns of prosperity and displacement, and gain a broader perspective on where newcomers reside and how long they stay. Are there characteristics of an "arrival city" as explained by Saunders?

A great resource for the city block and ten-square-blocks scale is Harmon's *You Are Here: NYC: Mapping the Soul of the City*, which has dozens of artist portraits of New York City neighborhoods. And for a fascinating citywide portrait, see Rebecca Solnit and Joshua Jelly-Schapiro's *Nonstop Metropolis: A New York City Atlas*. Both books are richly illustrated, and have many ideas that can be adapted for curriculum or used as travel guides. Solnit and Jelly-Schapiro have written similar books for New Orleans and San Francisco.

3. Migration and You: Personal Identity

I recommend this for all readers of this book, whether or not you are in an educational setting. The best way to understand human migration is to chart your own family history. You do this by speaking to as many family members as possible, asking them how much history they know, and tracing all the places they've lived. Given the dynamic demographic changes over the last two centuries, in most cases you will compile a surprisingly detailed and expansive map. See if you can focus on the migration stories, either from an original home country to a destination country or within a home country. These kinds of interviews are helpful in building understanding about the ubiquity of migration experiences and perhaps empathy for people who are currently migrating. At each step along the way, it's interesting to ask questions about livelihood and environment. How did your family members earn a living, and how did the ecological environment influence them?

As a second step, depending on available resources, I highly recommend one of the many available ancestry programs so you can trace your migration path back multiple generations. There is no better way to understand the extraordinary diversity of every human individual, and now more than ever, we need that perspective. Similarly, it's intriguing to research environmental and social conditions at each step along the migration path to better understand why people might have moved when they did.

These activities also raise important questions about who belongs in a place, how such decisions are made, and how depending on the migration life cycle of multiple generations, the length of a residency can be tenuous indeed. The personal stories are interesting in helping to understand your ethnic roots, the various displacements that are lodged in your ancestry, and how those histories impact your psyche. For example, my parents were only one generation removed from ghettos and pogroms, and although I grew up in a secure and safe setting, their experiences had a major impact on their inability to trust people who weren't family members or Jews, and impacted their personal confidence too. I have no doubt that those experiences were important aspects of their child-rearing practices and influenced me in all kinds of ways. I present this illustration not as psychological self-help but rather to lay the grounds for cultivating empathy for the current generation of migrants.

In the next chapter, I'll look at the idea of cosmopolitan bioregionalism and how to attend to the significance of place, while developing a fidelity to the ecological and cultural dynamics of your local environment as well as respecting the necessity of a cosmopolitan ideal. This discussion and educational process is crucial for understanding migration, and overcoming so many of the polarizations that divide us.

7

Cosmopolitan Bioregionalism

Maps, Patterns, and Places

Map-readers enter the world of the map, and in that process become changed by it. The map in the mind provides the grid points of cognition; and our new maps of distant and abstract landscapes promote new geometries of thought—new associations and therefore new ways of thinking about the world "out there"
—Stephen S. Hall, *Mapping the Next Millennium*

Each of us is an atlas of sorts, already knowing how to navigate some portion of the world, containing innumerable versions of place as experience and desire and fear, as route and landmark and memory.
—Rebecca Solnit and Joshua Jelly-Schapiro, *Nonstop Metropolis*

I can't remember the first time I laid eyes on a map. But I can tell you that I was happy to have one in my hands. By the time I was eight years old, I had a shoebox full of road maps. In those days, every gas station carried them and they were free. I studied those maps, played with them, and used them for all kinds of imaginative fantasies. I particularly enjoyed the maps of New York City, tracing the highways and routes around the metropolitan area. When I was nine years old, I made my own map of the bike circuits and roads within a five-mile radius of our home. I even put a gas station logo on the map, inventing a company called Shade. When our family took a western trip in 1962, my collection was greatly expanded and before long I had road maps for much of the United States.

In the late 1950s while digging through the bookshelves at my Aunt Helen's house, I discovered *Goode's World Atlas*. It was filled with maps that spanned the globe. They were different than the road maps I was accustomed to. They depicted topography. There was a huge section of world maps that were thematic, displaying the location of commodities and resources, weather and climate, and population trends and settlement patterns. Aunt Helen graciously let me keep the book, and I spent hours exploring the great big world.

As I grew older, and as information graphics and cartography became increasingly sophisticated, I learned about topographic maps, geologic maps, and maps of the heavens and oceans. I needed another huge box for an extensive collection of *National Geographic* maps. By the 1970s, *National Geographic* was publishing sumptuous thematic maps. In one of the many places I lived, I plastered an entire wall with those maps.

Now I wander through the shelves of my library to browse through several dozen atlases, artistic map books, and books that use maps to organize ideas, such as —*Earth from Above, Mind the Map, The Map as Art,* and *Cosmigraphics,* among many others. I am grateful that these wonderful books expand my geometries of thought, elucidating patterns and relationships that would be hard to ascertain if they were confined to text.[1]

In the late 1970s and 1980s as I was developing curriculum for graduate courses in environmental studies, I used my passion for maps to generate narratives around environmental values. One of my first "creative" assignments was the "sense-of-place" map. I still utilize this approach in my teaching. I ask students to construct maps (broadly conceived) that represent an environmental autobiography, including places they've lived or visited that had an important influence on their ecological identity. The wide range of artistic, conceptual, and narrative responses cultivates an enriching dialogue, encompassing a wonderful variety of backgrounds and experiences.[2]

These maps typically raise controversial questions about the political meaning of borders, boundaries, and territories. Who were the original inhabitants of a place, and who claims sovereignty? And they raise evocative questions about ecology and landscape, including the migration of peoples and species. They also raise epistemological questions. What does a map contain, and what does it leave out? I hope that the many mapping activities in this book, all of which are designed to help readers visualize complex relationships, lead to these kinds of questions too. When they do, they serve to enhance environmental awareness.

The bioregionalism movement was emerging at the same time. In 1981, the *Coevolution Quarterly* (the magazine extension of the *Whole Earth Catalog*) published a special issue on bioregions. The first page included the influential "Where You At?" quiz that challenged readers' knowledge of local natural history—"a self-scoring test on basic environmental perceptions of place." The inside cover contained a "Devolving Europe" map that featured dozens of small principalities reflecting home rule movements. The first essay, "Living by Life" by Jim Dodge, explored the philosophical foundations of bioregionalism, a motley combination of ecology, anarchism, and tolerance. As Dodge suggested, "Bioregionalism is hardly a new notion; it has been the animating cultural principle through 99 per cent of human history, and is at least as old as consciousness."[3] That is, for sure, a bold and provocative claim. Yet it still serves as a powerful ecological and political conscience for the environmental field. There is still substantial interest in these ideas, and mapping your bioregion, or whatever you wish to call it, is a powerful means for community empowerment. Doug Aberley's anthology *Boundaries of Home* initiated this approach, and inspired a vibrant and useful literature of practice.

In a previous work, *Bringing the Biosphere Home*, I explored the applicability of bioregionalism to global environmental change, especially given the ubiquity of migration and diaspora, relative transience of contemporary life, and dynamic changes in electronic communication, challenging how we conceive of time and space. I wrote a complementary essay, "Toward a Cosmopolitan Bioregionalism," in an anthology about bioregionalism that described bioregionalism as an aspiration seeking to integrate "ecological and cultural affiliations within the framework of a place-based sensibility, derived from landscape, ecosystem, watershed, indigenous culture, local community knowledge, environmental history, climate, and geography."[4] To affix cosmopolitan as a prefix suggests an emphasis on tolerance and dynamic cultural mixing, promoting an affiliation with global citizenship. Another way to think about the juxtaposition of these words is to recognize that bioregionalism stresses the ecological parameters of living in a place, while cosmopolitanism highlights how different cultures within that place live together and dwell on a common ground.

I present this retrospective not only because it places my own career and learning in perspective but also because it allows us to reflect on the environmental field more broadly—where has it come from and how it must proceed. Now, forty years after the sense-of-place maps and the first excitement about bioregionalism, do these ideas still enhance environmental learning? Where do these ideas fit on our collective cognitive map in a time of dynamic global environmental change, in an era when so many people are displaced and uprooted, during a time of megacities sprouting up all over the planet?

Now more than ever, it's crucial to explore the unfolding dynamics of the local and global—how place-based actions connect to global environmental challenges, and how global issues impact our local communities. I'll start with a reminder: there are over four thousand indigenous communities—First Nations—with distinct ecological and cultural traditions, all of which have legitimate claims to ancestral homelands. Globalization consolidates those territorial claims into 195 countries, whose territorial boundaries are still often contested. I'll elaborate on the concepts of borders and boundaries, from both ecological and political perspectives, stressing that sovereignty claims—adjudicating the challenges of tribes and territories—remains a powerful hold on personal as well as cultural identity. I'll suggest that the concept of place has always been fundamental to environmental learning, and bioregionalism remains a thoughtful, vibrant, visionary, and practical policy application of that perspective. Furthermore, I'll claim that global urban sustainability initiatives and the growing field of urban ecology represent a vital and influential movement, understanding the necessity of connecting the city to the biosphere. There's an emerging recognition that with one out of every twelve planetary citizens a migrant, refugee, or immigrant, a place-based perspective must entail a broader view, accommodating an enriching complex of planetary identity narratives. So many stories must be spoken and heard! Now more than ever, we require such narratives as well as educational strategies that welcome newcomers, promote tolerance and respect, allow for uncertainty, and facilitate adaptation and resilience while teaching us all how to live with difference.

The Original Bioregionalists

According to the UN, there are 370 million self-described "indigenous peoples" alive today. Generally defined as people whose community preexisted a larger nation-state that enveloped them, indigenous people comprise about 4,500 distinct cultures and speak as many different languages and dialects.
—Mark Dowie, *The Haida Gwaii Lesson*

To gain a deeper appreciation of the extraordinary diversity of peoples and cultures, scroll through the outstanding Wikipedia list of indigenous peoples.[5] Indigenous peoples represent approximately 5 percent of the world's total population and occupy about 20 percent of the planet's land surface. Please keep in mind that these distinct cultures (4,500) "exist within and straddle the borders of 75 of the United Nations 193 recognized countries."⁶ As Dowie reminds us, these cultures live in equally distinct habits and ecosystems.

Indigenous peoples live in tropical forests, boreal forests, deserts, and snow as well as on tundra, savannas, prairies, islands, and mountains, and occupy every remaining complex biotic community (or "biome") on the planet. They are stewards of about 80 percent of the world's remaining biological diversity and account for 90 percent of its cultural diversity.⁷

There's a significant research literature that indicates cultural diversity and biological diversity are mutually reinforcing, and the languages of indigenous cultures represent veritable encyclopedias of natural history, ecology, and environmental change knowledge.⁸ Dowie suggests that most indigenous peoples regard these ancestral habitats as nations, First Nations, and "would draw a very different map of the world than that found in modern atlases."⁹ You can get a glimpse of what such a world map might look like by viewing the interactive map of the world's indigenous peoples on the LandMark website, subtitled the Global Platform of Indigenous and Community Lands.¹⁰

Notice how these habitats (or territories) are juxtaposed within country boundaries. It's worthwhile to spend some time reconceptualizing borders and boundaries in this way as it helps you visualize the dramatic impacts of European colonization along with the stretch and reach of historic empires—the obsessive compulsion to satisfy the egotistical gratifications of territorial aggrandizement. There are many comprehensive histories of this process, and I need not reiterate the exploitation contained in those histories here. One of the best is by Charles S. Maier, *Once within Borders*, which magnificently traces global enclosures since 1500, and in the process reveals the complex relationship between power, wealth, territory, and belonging, informing any good conversation about the political dynamics of nationalism, and offering a sobering challenge to utopian notions of cosmopolitanism and/or bioregionalism. I'll look at some of those issues shortly. For now, let us remember the consequences of squeezing 4,500 distinct cultures into 193 countries, recognize the processes that led to such a powerful centralization, and at the very least raise consciousness about the dislocations, deracinations, and delegitimations that are still very much with us.¹¹

There's another angle here. I am in awe at the wondrous diversity of people and cultures that once roamed the planet, and still reside here, and how many are revitalizing their cultural practices and ecological knowledge. Dowie further reminds us that for "most of the millennia they endured, these small nations knew little of each other's existence," but now, via the internet and the very same globalization processes that threaten their demise, they have managed "to cobble together one of the most remarkable world-wide social movements in the history of mankind, a movement to reassert their rights to self-determination and claims to the territories they have traditionally occupied."[12]

Let's add another layer to these 370 million First Nations people and superimpose the 272 million (and growing) migrants roaming the planet. There may be some double counting here, as some percentage (such figures are unavailable) of migrants are further displaced indigenous people. As of October 2019, the world's population is 7.7 billion people. So First Nations people and migrants comprise approximately 8 percent of that number, or about 1 out of every 12 people on the planet. Where are these people going, and what will happen to them? What rights can they claim, resurrect, and reinstate? And to what place do they now belong?

We might describe First Nations cultures as the original bioregionalists, and the modern concept, mainly developed by contemporary environmentalists, is an adaptation of how most people have always lived on the planet. But since 1500, with the exception of a few outlying communities, it's been extremely difficult to sustain a bioregional approach to livelihood. The bioregional movement, with its notions of reinhabitation, is an attempt to revitalize what the world's First Nations have pretty much always known. As globalization and migration proceed apace, of what value can testimonies to place possibly be? Yet the appeal of place and its manifestation in bioregionalism still have much to offer, both as an approach to governance and locus for personal identity.

This is what I'll examine next, but before I do so, let me parse the concepts of borders, boundaries, territories, and ecosystems, and see where it leads us.

Borders, Boundaries, and Ecosystems

The mind wants boundaries.
It sees a mantle without shape, astir
with a great probe of life.
life surfacing, flexing, burgeoning fit to burst.
The mind wraps it the planet round.
one vast molecular swarm,
swollen, unconfined with life's first spring.
It is without bounds, says the mind,
and the mind demands beginnings, ends.
—Pete Hay, *Physick*

Despite the huge temporal and spatial differences involved, many boundary characteristics are similar. Boundaries are produced by environmental discontinuities, natural disturbance, and frequently human activity. Most are ephemeral in a century or millennium scale, but some persist. They are differentially permeable. They change internally. And they often move.
—Richard T. T. Forman, *Land Mosaics*

The words border and boundary contain a subtle distinction. "Border" refers to the edge or boundary of something, and is now commonly used to mean the "line separating two political or geographical areas, especially countries." "Boundary" refers to "a line that marks the limits of an area."[13] As far as I can tell, there's not a heck of a lot of difference between the two terms. In common usage, border is a political application of boundary.

To expand how we conceive of borders and relate the concept to environmental learning, it's enlightening to consider how ecologists think about boundaries. In one such formulation, "A Classification of Ecological Boundaries," five coauthors delineate no less than twenty-five attributes of ecological boundaries, organized according to how the boundary originated, how it's maintained, its spatial structure, its functions, and how it changes over time.[14]

J. Kolasa, a biologist, expands on this work to provide ecology researchers with consistent methodological frameworks for understanding this elusive concept. He distinguishes six "mixes of boundary-creating forces," including cultural, territorial, and range boundaries, and then "boundaries among associations of individuals caused by biotic interactions such as predation, mutualism, or competition, boundaries among communities and ecosystems, and boundaries among ecosystem complexes such as biomes."[15] Kolasa discusses these in detail while pointing out the multiple interconnections, permutations, and spatial or temporal layers in each case. Ecological boundaries are permeable, dynamic, and interconnected.

Switch gears for a moment, and check out a wonderful video that visually re-creates one thousand years of European border changes in three minutes.[16] The country boundaries or borders (what are they?) move so quickly that it's almost comical. At the very least, you recognize the utter ephemerality of national borders. Indeed, borders between nations are a manifestation of territorial imperatives, thus linking our obsession with them to evolutionary psychology as well as the persistence of tribes and territories. I don't wish to diminish their importance, or even necessity, as much as I want readers to recognize their broader ecological and historical context.

So what is a boundary? It's a way of making a distinction. Ecologists classify boundaries to organize their research, clarify scale, and develop a consistent methodology while recognizing the limitations of doing so. Communities and cultures delineate boundaries to mark differences, distinguishing themselves for purposes of identity, sovereignty, and social organization. Migration processes challenge those distinctions. Migrants move between borders, raising questions of belonging, identity, and autonomy.

A border is a type of boundary, organized for the purposes of political distinctions—bringing regional or international order through towns, cities, counties, states, and nations, or whatever they are called where you happen to live. Borders are the source of multiple forms of contention. Do we agree on where they should be? Who gets to pass through them? Who has sovereignty over border zones? Think about the controversies that surround the perceived legitimacy of the borders of nation-states. China is currently constructing artificial islands in the South China Sea. Should that challenge the sovereignty of surrounding nations? How many miles should a nation's territorial sea boundaries extend? Should we build a wall to solidify a demarcation zone? What do border walls signify?

Let's return to the natural world for a moment and consider the ecological challenge of invasive species—another type (I suppose) of unwanted migrants. Some invasive species arrive by virtue of the naive intentions of colonists. Who would have known that starlings would become so prevalent in North America? We'd rather not spend countless hours trying to eliminate Japanese knotweed and Oriental bittersweet from our garden areas. And exactly how did those European green crabs migrate so rapidly down the coast of Maine? How then can we understand the ecological and human parameters of permeable borders and thriving ecosystems? Are there ecological and cultural criteria for better understanding the relationship between migration and boundaries?

Let's return to the political controversy that surrounds the concept of open borders. This is an important and fascinating dialogue. Have a look at the comprehensive Open Borders website to get a sense of the depth of the controversy and different interpretations that emerge.[17] There's a subsection that describes three foundations for a moral case supporting open borders, drawn from libertarian, utilitarian, and egalitarian positions. These approaches cover every conceivable wing and partition of the political spectrum. Free traders, typically strong advocates of so-called free markets, tend to favor open borders, although the recent racially tinged controversies over migrants has muddied the waters. Progressive-minded or egalitarian-oriented thinkers often oppose free trade because they worry that it favors powerful multinational corporate interests over local workers, yet they tend to be more favorable to liberal immigration policies.

We seem to be in a transition zone, using the legacies of imperial territorial approaches to sort out the dynamic challenges of human and ecological migration. Our challenge arises from our inability to internalize the global transformations that accompany the Anthropocene, especially the environmental dislocations as well as conceptual upheaval catalyzed by an expanded concept of networks. Ownership of networks may be another form of territorial contention. The very idea of territory is undergoing transformation as the interconnections between the global economy and ecology demand entirely new approaches to these issues.

What's missing from much of the popular discussion around borders is an integration of ecological criteria and political philosophy, particularly as applied to transboundary issues, which now must include human migration and First Nations as central to the conversation. Once political philosophy enters the picture, we encounter property, territory, and sovereignty—human constructions superimposed on ecological landscapes.

Tribes, Territories, and Sovereignty

We cannot renounce territory, even if it does less for us. Diasporas, multinationality, and advanced capitalism have all attenuated its significance; however, its sentimental importance has not thereby weakened, at least not yet.
—Charles S. Maier, *Once within Borders*

Ethnie always possess ties to a particular locus or territory, which they call their "own." They may well reside in that territory; or the association with it may be just a potent memory. An *ethnie* need not be in physical possession of "its" territory, what matters is that it has a symbolic geographical centre, a sacred habitat, a "homeland," to which it may symbolically return, even when its members are scattered across the globe and have lost their homeland centuries ago.
—Anthony D. Smith, *The Ethnic Origins of Nations*

If you scroll through the Wikipedia list of contemporary ethnic groups, you will likely be amazed at the sheer number of them: at least several hundred distinct groups based on cultural heritage, ancestry, history, homeland, and language, and then at least another thousand subgroups.[18] Consider that a list such as this is merely a snapshot of a moment in historical time. Like the history of human migration, the differentiation of tribes along with their cultural arrangements, ecological habitats, and perception of homeland begets a multitude of territorial possibilities. Contrast a list like this with the historical shifting of territorial borders, rise and fall of empires, and all the contested claims of sovereignty, and you have a complex mess: hundreds, if not thousands, of groups that lay claim to an original homeland, typically a symbolic place given the historical shifting of peoples and borders.

Globalization represents an integrating potential, weaving these numerous groups into comprehensive networks of cultural and economic exchange. The enormously complex nation-states (China, Russia, India, Brazil, Indonesia, and the United States, among others) must somehow assimilate these groups into a coherent sense of nationhood while managing the unprecedented flows of people, resources, and ideas between their already-cumbersome administrative systems. Attempts to coordinate this—such as the European Union—have their supporters and detractors, as some claim the efficiencies and benefits of integration, and others worry about the unfreedom of sequestered autonomy. Superficially, this appears as a duality—the populists verses the cosmopolites—but the reality is more of a dialectic. Populists communicate through shared international spaces (the internet) and enjoy many of the benefits of globalization, and cosmopolites may celebrate their ethnic origins or other affiliations as measures of belonging.

The reality is that few ethnic groups reside in their symbolic, sacred homeland. Yet the myths of distinction, differentiation, and belonging still carry enormous appeal, not only for the purposes of personal and cultural identity, but to organize politically as a bulwark against the anonymity and exploitation often accompanying globalization. Given the circles of overlapping territorial claims, it is almost impossible to conceive of equitable resolutions based on the idea of an ancestral homeland. As Dowie explains in his case study of the Haida Gwaii, however, imaginative arrangements, organized around inclusion, participation, redistribution, and cooperation, enable the prospect of fluid sovereignties and mutual cultural respect.

What a great irony that the globalization process makes such new coalitions and arrangements possible. Dowie observes that "as the world shrank due to telecommunications and transportation, inhabitants of the most remote villages began to discover that they were not alone. There were others like them on almost every continent—people with unique dialects, diets, cultures, and cosmologies who were at best misunderstood, at worst oppressed by the dominant nationalities that surrounded and subsumed them."[19]

It's a unique trademark of the Anthropocene that First Nations can organize this emerging awareness by virtue of dynamic websites. Here's the other great irony. Within majority populations, there are supremacist groups that advocate extreme measures to solidify their hold on the perceived privileges they believe are evaporating before their eyes. They use the same internet communication techniques to build solidarity, reinvigorate their notions of supremacy, and build idealized communities based on nostalgia for a mythologized past. Hence we are living in a global nationalist revival, clinging to pathologies of supremacy, catalyzed by the very technologies that allow for their reemergence.

How can any of these conflicting claims or clashing narratives be peacefully resolved? There's much discussion about ways of mitigating modern tribal differences—from Greene's "six rules for modern herders" in his book *Moral Tribes* to Sapolsky's advice for transcending the intrinsic us versus them dilemmas that lie deep within every human brain.[20] Indeed, there are many emerging processes for engendering compassion, perspective taking, conflict resolution, and cultural exchange. The more we develop these techniques and educate ourselves about them, the more skills we have at our disposal.

I'd like to elaborate on a parallel course, based on the wisdom of bioregionalism and the necessity of a cosmopolitan ideal. Territory still matters because life has meaning when people have a place they can call home, and when that home connects to a broader cultural community. The bioregional perspective claims that the place we call home also includes species, habitats, and landscapes. There is no human flourishing without this awareness, this sense of belonging, or what environmental learning describes as a sense of place. Yet sense of place can no longer be consummated by residing in an original homeland. There are too many people, too many claims, and too much dynamic change for that. It must now be incorporated within a cosmopolitan ideal—an approach that allows us to construct and navigate fluid sovereignties, adaptive communities, diverse cultural perspectives, shifting ecological boundaries, and multiple narratives of shared belonging.

Why Place Still Matters

Place is a fluid concept that embodies identity within landscape and geography. Environmental thinkers have long asserted that place is a common ground for deepening an awareness of ecological relationships within a community. Place refers to where and how people live in a community, suggesting that dwelling implies care, responsibility, and affiliation, as influenced by home, habitats, and landscapes.

Cultivating a sense of place has appeal because it conveys feelings of rootedness and stability in a world of dynamic change. Fidelity to place is a response to globalization and all that it entails: people on the move, economic and cultural trends that sweep through communities, and the formidable pace of environmental change. By achieving a sense of place, you learn to identify with the place where you live. You become familiar and intimate with your local surroundings. Hopefully you exercise citizenship by taking responsibility for that place's quality of life, and taking action to care for it in a tangible and meaningful way.

Citizenship means you learn to work with your neighbors. You have a better understanding of their needs and concerns. You come to grips with the reality that you all live in your place together. You find consensus in the midst of controversy, listening to multiple voices and engaging with people who have ideas different than your own. Common ground links place, landscape, and decision making. Through community conversation, you deepen what you hold in common, knowing some differences may always remain.

Pride in place has deep roots in European American culture, reflecting a long tradition of associating virtue with ancestry and landscape. Such presumed virtue has a shadow side as well, though, as pride in place often comes before a fall. Pride in place can be a convenient excuse for exclusion, discrimination, and superiority, taking the soft form of NIMBY, or the hard form of racism and anti-immigration sentiment. Nationalism is pride in place translated to the scale of the nation-state.

What value does sense of place retain in a time of dynamic human mobility? We've already covered the challenge of welcoming migrants and immigrants, or respecting multiple claims to sovereignty. And millions of people move—rich and poor alike—to follow the trail of international capital, which is more interested in profit than fidelity to place. How, too, do you promote a sense of place in dynamic metropolitan areas, where the flows of mobility and capital potentially overwhelm neighborhoods?

Place matters most when communities consider the reality of these flows, both urban and rural, and when communities accommodate the shifting nature of people and capital, finding ways to promote inclusion, fidelity, affiliation, tolerance, and a deeper understanding of the ecological processes that allow for human flourishing. In an earlier section, "Pacific Northwest Change Makers," I described community projects that are accomplishing exactly that, and suggested that there are many hundreds and even thousands of such projects around the world.

I believe that affiliation to place, as linked to promoting ecological awareness, respect for diversity, inclusive decision making, and a commitment to equity, is a powerful global movement. There are countless projects that affirm these values and have the results to show for it. They are grass roots, multigenerational, multiracial, and consensus oriented. They build revenue through a pastiche of approaches, such as crowdfunding, entrepreneurship, impact investment, philanthropy, volunteerism, and other forms of community and civic support. They understand that there are challenging obstacles so they build coalitions of support and mentoring. For many of these projects, place provides an ecological as well as psychic matrix, promoting a conceivable scale, coherent objectives, and a desire to build lasting community.

The best way to explore the depth of this movement is to go out (wherever you may live) and see for yourself. To further inspire those trips, it's enlightening and encouraging to do a web sweep of place-based projects, so you can appreciate the scope, variety, and potential of what's possible. Try some of these web locations to get started.

Urban blogs such as Next City (https://nextcity.org) and the Nature of Cities (https://www.thenatureofcities.com) cover dozens of innovative urban community projects. The Daily Yonder (http://www.dailyyonder.com), a communications wing of the Center for Rural Strategies (https://www.ruralstrategies.org), looks at place-based rural initiatives. The Project for Public Spaces (https://www.pps.org) implements "creative placemaking" strategies in multiple venues. If you search out the term "creative placemaking," you'll find dozens of vital projects in communities around the world. And there's a growing literature regarding placemaking among migrant communities. See especially Simon Pemberton and Jenny Phillimore, "Migrant Place-Making in Super- Diverse Neighbourhoods: Moving beyond Ethno-National Approaches." Many projects use a place-based approach to coordinate action around global challenges. See especially the portfolio of projects organized by Paul Hawken's "Drawdown" network (https://www.drawdown.org)—local projects designed to mitigate climate change.

So yes, place still does matter, and now more than ever, particularly when it serves to integrate diverse communities in changing landscapes. A place-based approach is the conceptual foundation for bioregionalism. Let's look next at why bioregionalism is still an excellent rubric for civic responsibility and political coordination.

Why Bioregionalism Still Matters

Stewardship means, for most of us, find your place on the planet, dig in, and take responsibility from there—the tiresome but tangible work of school boards, county supervisors, local foresters—local politics. Even while holding in mind the largest scale of potential change. Get a sense of workable territory, learn about it, and start acting point by point.
—Gary Snyder, *Turtle Island*

Bioregionalism is a body of thought and related practice that has evolved in response to the challenge of reconnecting socially-just human cultures in a sustainable manner to the region-scale ecosystems within which they are irrevocably embedded.
—Doug Aberley, *Boundaries of Home*

Snyder's "Four Changes," a beautiful short essay first published in 1969, served as an inspirational handbook for an entire generation of environmental learners. Thirty years later, Aberley defined bioregionalism while surveying the intellectual origins and practical applications of the term, providing an empowering focus for hundreds of grassroots efforts. In that same anthology, *Bioregionalism* edited by Michael Vincent McGinnis, I wrote an essay, "Cosmopolitan Bioregionalism," that presented approaches for linking the bioregional philosophy to global environmental change.

Although bioregionalism is certainly not a household word, and in large measure has been overshadowed by the more recent emphasis on sustainability, I have no doubt that five decades of environmental activism reflects an internalization of its basic concepts. For example, the recent anthology by Jennifer Chirico and Gregory S. Farley, *Thinking like an Island: Navigating a Sustainable Future in Hawai'i*, essentially provides a bioregional strategy for sustainable community planning, although the word bioregionalism never appears. The campus sustainability movement and most of its community-based applications—from local food to slow money—utilize bioregional, place-based principles as intrinsic to their efforts.

Consider, too, some of the introductory comments in a recent report by Peter Forbes, *Finding Balance at the Speed of Trust: The Story of Southeast Sustainable Partnerships*. "Now, this landscape is shaped by a new generation, children of The Timber Wars and of lost generations, yearning to create a new era of balance," notes Forbes. "As the sawdust has settled, the original tribal communities of Hoonah, Kake, Yakutat, Angoon, Craig, Klawock, Hydaburg, Sitka, and Kasaan are having a cultural renaissance even though there are few jobs, the timber money is all but gone, the forest around them is yet recovering, and the cost of living is high just as the population is low. The people still rely on the forests and the streams, and salmon is still part of their DNA. Can there be an opportunity to live whole, modern lives without scarring the land?"[21]

Consider this alongside the proliferation of bioregional ideas at the core of ecological urbanism—an approach to cities that emphasizes green living, sustainable planning, local knowledge, and collaborative visioning. Or through Benjamin Barber's vision: "If mayors ruled the world, the more than 3.5 billion people (over half of the world's population) who are urban dwellers and the many more in the exurban neighborhoods beyond could participate locally and cooperate globally at the same time—a miracle of civic 'glocality' promising pragmatism instead of politics, innovation rather than ideology, and solutions in place of sovereignty."[22]

The word bioregionalism may no longer be trendy, but I think we should acknowledge its influence and strongly consider its resurrection. Why? Because it makes clear the essence of environmental learning: that ecological and evolutionary patterns must be the foundation of political decision making, cultural practices, community planning, and educational approaches.

The challenge is how to make a concept that can be too easily dismissed as utopian, environmental, and rural relevant for an increasingly urban world, characterized by the dynamic flow of people and capital. It's too easy to separate as well as stereotype the local and global, the city and country, the settled and wild, home and away, and us versus them.

The polarizations that are constantly cited in the media and exploited by hand-wringing politicians who benefit from their reiteration are easy to bemoan. But it's far better to cultivate the narrative inspiration of people from different backgrounds working together, bridging the gaps of separation, and doing so peacefully, strategically, and honestly. When this occurs, the stereotypes may not disappear, but they fade into the background with the common interest of improving the life of a community.

I have a hunch. Take away the constant media attention given to polarizing factions, extreme rhetoric, and ephemeral tweets masquerading as public policy. Take away the massive amounts of money coming from billionaires who are peddling inappropriate influence. And instead direct that money and attention to the thousands of grassroots, place-based, community-building, bioregionally oriented projects that you can find just about anywhere on the planet, and you'll perceive a different world. Replace the hand-wringing with hope, corruption with integrity, election marketplace with face-to-face community meetings, and you'll be amazed at what you can accomplish.

In my view, the essence of a revitalized bioregionalism is permeable borders, thriving ecosystems, and community-based democracy. By permeable borders, I don't mean a free-for-all but rather an understanding that the same interconnectedness that allows you to use the internet also brings displaced persons to your community, and there is a relationship between welcoming strangers and the quality of a community. By thriving ecosystems, I refer to an understanding that every consumer action ultimately has an ecological impact, human flourishing cannot be measured in economic terms alone, and human communities require healthy habitats. By community-based democracy, I refer to an understanding that local knowledge builds community wisdom, and we need investment in the local institutions that cultivate such knowledge. That would be a far better use of money than squandering it on election advertisements.

And finally, I refer to the necessity of internalizing a cosmopolitan ideal: to see diverse cultures as enriching, as engines for innovation, and the impetus for deeper learning; hence a connective bioregionalism. We have much to learn about bioregionalism, and its full potential is as yet unrequited. Now perhaps in the early days of the Anthropocene, with the reality of rapid global environmental change, its time has finally come.

The Biosphere and the City

But inside of New York another way of thinking is emerging, a new set of ideas and beliefs that do not depend on disaster to correct our course and instead imagines a future where humanity embraces, rather than disdains, our connection to the natural world. Many New Yorkers celebrate the nature of their city and seek to understand the city's place in nature. They see their city as an ecosystem and recognize that like any good ecosystem the city has cycles, flows, interconnections, and mechanisms for self-correction.
—Eric W. Sanderson, *Mannahatta*

When I lived in the New York metropolitan area in the late 1960s and early 1970s, I had several recurrent dreams. In one of them I was floating in New York Harbor, in a time before human settlement, gazing longingly at the forests and meadows of Manhattan Island, cruising through the various watercourses and islands that are now New York City. In another, I would enter a doorway in some part of the city and on the other side find a vast wild landscape, with hills and streams. In the dream, I felt as if it was a place I visited many times, but in my waking hours I could never find that secret door. In the third dream, Long Island, where I spent much of my childhood, had an entirely new topography, with an exquisite mountain range in the middle of the island (where there is in fact a terminal moraine).

Many years later, in 2009, I discovered the exceptional book *Mannahatta*. Sanderson and his team of researchers at the Wildlife Conservation Society, using historical maps, paleoecological data, computer graphics, and various simulations, reconstructed New York City before Anglo settlement, re-creating the landscape ecology during the time of the Lanape, the original bioregionalists of the island. The Lanape called the island Mannahatta, or the "island of many hills." Sanderson and his team stunningly display (in maps and computer graphics) the complex and diverse ecosystems of the island.

Mannahatta had more ecological communities per acre than Yellowstone, more native plant species per acre than Yosemite, and more birds than the Great Smoky Mountains National Park.[23]

They assert that "if Mannahatta existed today as it did then, it would be a national park—it would be the crowning glory of American national parks." Yet this glorious vision of a harmonious Mannahatta is not merely a utopian dream. Sanderson's goal is to use the original landscape (at least as it existed in the sixteenth century) as a template for a revitalized approach to urban ecology and regional planning. He recognizes the astounding cultural diversity, its unsurpassed intellectual capital, its dynamic flow of people and ideas, "what might be the most exciting and stimulating city that has ever existed, providing home and satisfaction to extraordinary numbers of people, with previously unknown levels of knowledge, power, and freedom, to make choices about the kind of lives they want to lead." The challenge is to balance this dynamic urban complex with Mannahatta, "the best of nature's abundance and resilience."[24] The book concludes with a portrait of what Manhattan could become in 2409, if ecological wisdom is blended with innovative urban planning. Surely Sanderson's vision is a bioregional one, using ecological criteria as the foundation for regional planning, governance, and urban policy. I'm especially taken with how he celebrates both the ecological and cultural diversity of one of the densest cities in North America.

Flip through the wonderful anthology *Ecological Urbanism*, edited by Mohsen Mostafavi and Gareth Doherty, and you will see the bioregional vision of Mannahatta come to life in projects and places all over the world—New Waterscapes for Singapore, Natura Garden in Mexico City, One Airport Square in Accra, Ghana, the Uppsala Power Plant in Sweden, and literally hundreds of projects, innovations, experiments, and designs, at various scales, in every conceivable type of city and ecological region. The book also reflects the fascinating new theoretical approaches that integrate multiple disciplines, all oriented around ecological approaches to city and regional planning. This work builds on many generations of ecologically oriented planning.

Thankfully, with the recognition of the dynamic urbanization of the planet, there's a proliferation of theory and research in the still-emerging field of urban ecology. One of the leaders in that research is Alberti (who I introduced in chapter 5), whose book *Cities That Think like Planets* is a remarkable synthesis of urban ecology methods, approaches, and research possibilities. I particularly appreciate how Alberti places urban ecology in the context of the biosphere, writing that "urban ecology faces a significant challenge: To position itself in the context of planetary change and to understand the role that cities play in the evolution of Earth."[25]

In acknowledging this, she further emphasizes that "a science of cities as novel ecosystems has yet to be developed." Alberti proposes that we conceive of cities as hybrid ecosystems, which are "characterized by complex interactions (e.g., heterogeneous, nonlinear, multiple scale), emergent properties (e.g., patterns, processes, feedback), multiple equilibria (e.g., regime shifts), and the capacity for innovation."[26]

In elaborating on that last phrase, "capacity for innovation," Alberti reiterates that urban regions are the major drivers of global environmental change and we should now conceive of megacities as networked megaregions, characterized by connectedness and interdependence, supplying "new opportunities for both technological and social innovation."

Increased connections among people and places that occur across scales promote faster communication and learning. Parallel to structural and functional changes, emerging megaregions represent a shift in governance from formal centralized systems to hybrid, multiscale, cross-regional network structures that include both formal and informal institutions.

The emergence of new interactions and feedback across distant places as well as between local and global processes to megaregions offers unique opportunities to experiment with novel institutional frameworks. Alberti suggests that heterogeneous and modular systems are more capable of planning options that incorporate complexity, resilience, uncertainty, adaptation, and transformation.[27]

These four works taken together—*Mannahatta*, *Ecological Urbanism*, *Cities That Think like Planets*, and *If Mayors Ruled the World* (see the previous section)—provide a compelling vision of how environmental learning and ecological knowledge can be the foundation for urban planning. While they all are manifestly aware of the overwhelming challenges we face, they supply evidence that there's a global movement of emerging, innovative, ecology-based policies and practices that allow urban and rural dwellers alike to connect place to planet, participatory democracy with ecological planning, and community autonomy with complex networks.

But even if we see our way to ecological solutions, there are still challenging social and political issues when people with different cultural backgrounds and values interact, when tribes and territories mingle, and when the dynamic flow of displaced people enters your community. Just as there are innovative thinkers working on the city and biosphere, there's a body of research and practice that helps communities better understand how to encounter difference. That, too, is a prerequisite for cosmopolitan bioregionalism, so I'll review some of those efforts next.

Contact Zones: Living with Difference

How are people from different cultures, different backgrounds, with different languages, different religious beliefs, produced by different and highly uneven histories, but who find themselves either directly connected because they have to make a life together, or digitally connected because they occupy the same symbolic worlds—how are they to make some sort of common life together without retreating into warring tribes, eating one another, or insisting that other people must look exactly like you, behave exactly like you, think exactly like you?
—Stuart Hall, "Living with Difference"

If bioregionalism is to thrive as a political strategy, it must address this vexing question. The "living with difference" challenge is prevalent in both everyday encounters as well as larger-scale political issues. It cannot be wished away or solved by legislating separation or integration, although such measures are frequently considered and implemented. Globalization catalyzes proximity—entailing cohabitation or misunderstanding, equity or exploitation, and innovation or prejudice. Interconnectedness may be essential for bioregional policy coordination, but it takes courage to open yourself up to engaging with people who are different than you. And what we've learned thus far about social media is that people tend to communicate with those who are most like themselves.

There's a copious and instructive body of literature that analyzes these issues. One of the best literature reviews is an essay by Gill Valentine, "Living with Difference: Reflections on Geographies of Encounter." She surveys the sociological and geographic literature since 1950, and describes many of the studies covering concepts such as the contact hypothesis, parallel lives, spaces of interdependence, factors that contribute to cultural destabilization, and whether it's possible to "scale up" a politics of connectivity. Ultimately she suggests that it's difficult to overcome layers of historical prejudice and difference matters, concluding that to constructively address these challenges, scholars must focus on "sociospatial inequalities and the insecurities they breed," while "trying to understand the complex and intersecting ways in which power operates."[28]

Another important work is *Encountering Difference* by Robin Cohen and Olivia Sheringham. They identify five forms of social identity, or what they call "identity trajectories," that arose from "globalization, international migration and other rapid social changes." The coauthors offer a Venn diagram that portrays these identity trajectories as an overlapping matrix of possibilities. To paraphrase from a longer passage, these are "a reaffirmation of felt loyalties to *sub-national entities* like clan, tribes, ethnicity, locality, or language group, a revival of *nationalism*, a recasting of *diasporic identities*, a linking and blending with other groups through a process of *creolization and hybridization*, and the development of a universal spirit that transcends any particularities and simply stresses the quality of being human, namely the *cosmopolitan* possibility."[29]

Cohen and Sheringham are wise to suggest these five identity trajectories—subnational, nationalism, diaspora, creolization, and cosmopolitan—are interconnected, as surely we all can identify these belonging patterns in our psyches. I'll look at the educational ramifications of these possibilities in the concluding section of this chapter.

Most of their book is devoted to the idea of creolization, which they describe as the "mixture and continuing admixture of peoples, languages, and cultures." They argue that both creolization and diasporic identities potentially transcend the more limiting and typically exclusive expressions of identity that emerge with "forceful assertions of nationalism or religious certainties, and challenge the solidities of closed ethnic and racial categories."[30]

When creolization occurs, participants select particular elements from incoming or inherited cultures, endow them with meanings different from those they possessed in the original culture, and then creatively merge them to create totally new varieties that supersede the prior forms. Creolization thus evokes a "here and now" sensibility that erodes old roots, and stresses fresh and creative beginnings in a novel place of identification.[31]

They conclude that "diaspora and creolization are crucial concepts for thinking about identity in our dynamic, ever more interconnected world: a world of movement, a world of migrants, and a world of 'relation.'"[32] Most of the book is a review of several years of research in communities that embody creolization and an assessment of the cultural practices that enhance these possibilities. They discuss "contact zones" where such processes are most likely to occur, both historically and in the contemporary era, like islands and plantations, port cities, and super-diverse cities. They are always aware of power differentials and colonization histories in describing these contact zones, as similar to Valentine, it's crucial to understand that inequalities and insecurities have histories. Most important, their experience indicates that specific cultural practices are constructive venues for creolization, such as music, carnival and celebration, and food.

For educational purposes, the concept of contact zones has great potential. The term was first used by Mary Lois Pratt, a literary scholar, as "social spaces where cultures meet, clash, and grapple with each other often in contexts of highly asymmetrical relations of power, such as colonialism, slavery, or their aftermaths as they are lived out in many parts of the world today."[33] The term is now widely used among multicultural researchers, as it provides both a geographic and political metaphor for the interaction between different groups of people. It's also used among scholars who wish to facilitate interactions in multigenerational settings, so we now have multigenerational contact zones, or "spatial focal points for different generations to meet, interact, build relationships (e.g., trust and friendships), and, if desired, work together to address issues of local concern. They can be found in all types of community settings including schools, parks, taverns, reading rooms, clubhouses, museums, community gardens, environmental education centers, and multi-service community centers."[34]

I prefer the latter definition as an application for environmental learning, as it broadens the locations of contact as well as the types of discussions, exchanges, and constructive relationships that might develop. Really, any place where you encounter difference is potentially a contact zone, and these will inevitably be places for both cooperation, conflict, and tension. Many educators, social workers, and refugee organizations, however, strive to develop contact protocols to allow for such tensions to emerge as constructively as possible.

Let's pay attention to narratives that provide examples of constructive encounters in contact zones. We surely know enough about what might go wrong. For an inspiring, sobering, and emotional peek at one example, have a look at the remarkable book by Helen Thorpe, *The Newcomers*. Thorpe spent a year in the most rudimentary English-language acquisition class at Denver South High School in Colorado. This exemplary school describes itself as a place "where the world gathers to excel." The teacher, Mr. Williams, works with twenty-two students who arrive at various times during the school year, speak fourteen different languages, and come from exceptionally difficult circumstances. Here is a global mix of refugees, migrants, and immigrants who are fortunately enrolled in a school that respects their journeys, and does whatever it can to help them adjust to life in the United States. It's alternately wonderful and heartbreaking as the stories of these teenagers are slowly revealed over the course of the book and school year. There is no better illustration of what it means to encounter difference, how to deal with it constructively, and how to confront both the enrichment and tension that follows. Surely there are many other instances, in places all over the world, where contact zones become retreats for constructive learning and safe passage.

These students embody the five identity trajectories as identified above by Cohen and Sherington, and as they mature, they may well experience all of them. How can they do so inclusively, constructively, and as a way to learn about their place in the world? This question is relevant for all citizens in the Anthropocene, and clearly a foundation of how we all come to know the world. The school is one example of a contact zone. It's a place where teenagers with dramatically different backgrounds must learn to communicate in a new language, share whatever experiences they are capable of expressing, and build the skills of coping, learning, and hopefully succeeding after being uprooted multiple times in their young lives.

So What Is Cosmopolitan Bioregionalism?

The challenge, then, is to take minds and hearts formed over the long millennia of living in local groups and equip them with ideas and institutions that will allow us to live together as the global tribe we have become. . . .
There's a sense in which cosmopolitanism is the name not of the solution but of the challenge.
—Kwame Anthony Appiah, *Cosmopolitanism*

Cosmopolitan is a challenging and controversial word, conjuring contrasting visions. You might envision diverse groupings of people living together in world cities of mobility and opportunity, enriching each other by sharing stories of cultural traditions, exchanging dynamic ideas, and contributing to waves of innovation and entrepreneurship. Or you might envision the homogenization of values, a relativistic surfeit of secular consumerism, eviscerating the hard-fought legacy of national sovereignty, and cultural integrity. It's understandable how global migration contributes to both visions, and how easily these perspectives become polarized. Us versus them dichotomies fall prey to these polarizations. It's universalists versus particularists, cosmopolitans versus nationalists, free spirits versus traditionalists, globalists versus localists, or some unique mixture of all these perspectives. Is globalization our savior or downfall?

The reality is more nuanced and harder to clarify. That's why you have proponents across the entire spectrum of these potential responses, and it's almost impossible to classify them in traditional political terms. It's easy to see how migration and displacement, and the threats and opportunities they pose, are so difficult to discuss, especially when emotion trumps reason. Borders become symbols for all these conflicting values, impulses, emotional responses, challenges, and solutions. Should our borders be open or closed, rigid or permeable, walled or freely passable?

What happens when you put a fence around something? The border marks a territory, a distinction between places. Throughout the long waves of human history, borders pronounce enclosure, sovereignty, governance, and the rule of law. Yet from an ecological perspective, a border is ephemeral, a superficial boundary on the landscape, and form of organization that turns the layers and flows of the ecosystem into human territory. Such a juxtaposition is conceptually disorienting.

During the long history of life on earth, the migration of species unfolds according to ecological contingencies and evolutionary dynamics, and the same is true of human movement. Perhaps that is the ultimate ground of freedom. Can we establish criteria and protocol for borders that respect all these factors: human sovereignty, ecological flow, and the freedom to exchange ideas and cultural perspectives? Can we translate this into a cosmopolitan ideal while also respecting pride of place, the necessity of territory, longing for belonging, and inevitability of borders? Might we think of this blend as an emerging dialectic rather than intractable differences?

Let's reconsider this as the challenge of cosmopolitan bioregionalism. How can diverse communities, organized around fidelity to ecological place, cultural tradition, unique histories, and strong values, find ways to connect with each other and welcome newcomers, while building tolerance as well as respect for traditions and backgrounds that are different from their own? In some cases, these challenges emerge within a few city blocks. Sometimes they reflect the connection between communities, and those connections may be geographic or electronic.

And let's review the reality of global environmental change in the Anthropocene. One out of every twelve people on the planet is either a citizen of a First Nation, a migrant, refugee, or immigrant, or some combination of any of these possibilities. What we know about climate change, habitat destruction, and environmental disasters is that these problems are getting worse, and the number of uprooted people is on a steep upward trajectory. This uprootedness is not only an international challenge—people moving between borders—but also occurs in domestic settings. Think about the uprootedness after a California wildfire, hurricane that sweeps through Houston, earthquake in Mexico, or tsunami. Environmental change is exacerbated in the Anthropocene. It's a fact of life, and someday it might happen to you.

In *The Mushroom at the End of the World*, Tsing contends that disturbance ecologies, now prevalent in the Anthropocene, represent transition zones, places where ecological and cultural change are the norm. As an anthropologist, she wonders what it will take to survive or even thrive in such conditions. She advocates that we cultivate "precarity," or the condition of being both open and vulnerable to others. And she explains how a fear of contamination, based on a misunderstanding of the other, restricts adaptive learning. As Tsing observes, "Unpredictable encounters transform us; we are not in control, even of ourselves. Unable to rely on a stable structure of community, we are thrown into shifting assemblages, which remake us as well as others."[35]

Tsing suggests that we learn to live in these disturbance regimes, and we do so by embracing precarity. By looking deeply into our own vulnerability, we gain compassion and perspective for others, we're more likely to adjust and adapt to changing circumstances, and we are also more likely to perceive patterns of unintentional coordination, by which she means possibilities of contamination, convergence, and collaboration.[36] She argues that such a state of mind can revitalize political economy and environmental studies, as we can reinvent how we learn about environmental change and engage with conditions of uncertainty.

Is this, I wonder, also a precondition for the cosmopolitan ideal: the ability to engage constructively with disturbance, adapt to conditions of uncertainty, better understand the ecological and political dynamics that accompany environmental change, and prepare learners and citizens alike to deal with the reality of life in the Anthropocene?

Recall Cohen and Sherington's identity trajectories covered in the previous section—subnational, nationalism, diaspora, creolization, and cosmopolitan—and their suggestion that we view these as an interactive Venn diagram. Can a migrant, who's a member of a diaspora, maintain an emotional attachment to their ethnic group, place of origin, and new community? Certainly there are moments of divided loyalty, but they often result from polarization and political manipulation. Why can't people hold multiple local and global identities, and find ways to become what Ursula Heise describes in *Sense of Place and Sense of Planet* as ecocosmopolitans?

So what is cosmopolitan bioregionalism? Place still matters. Bioregionalism still matters. But both concepts, grounded in the local, must entail a global vision, and a way to understand the complicated connections between places and planet. Most important, this vision must offer compelling narratives—perspectives that enhance meaning in a time of dramatic uncertainty.

Let's start with the imagination. What are the narratives that describe not just your home place but all the connections—cultural, ecological, political, and historical—that link your place to the planet too? Which metaphors are most appropriate? Which stories do you tell your children, and how do you share them with your neighbors? On the most basic level, this sharing of stories involves common courtesy. At the very least, as Appiah asserts, let's discuss our obligation to strangers, obligation not to carry the burdens of the world alone, and most important, obligation to engage in discussion. And as Appiah also underscores, sharing stories and experiences is more about developing understanding along with respect than it is about assimilation or even approval. "You can be genuinely engaged with the ways of other societies without approving, let alone adopting them."[37] We can get to know each other better, without having to live similarly.

On a broader conceptual level, cosmopolitan bioregionalism requires narratives that connect the local to the global, and does so with compelling metaphors, responsive to global environmental change, derived from sound research linked to personal experience, with a willingness to experiment and explore. There are templates for this challenge. As Dowie's case study of the Haida Gwaii demonstrates, shared sovereignty emerges from sound ecosystem management, an understanding of diverse cultural perspectives, and a willingness to consider historical claims to ancestral ground. Haida Gwaii is a superb example of cosmopolitan bioregionalism: a First Nations culture with a ten-thousand-year history connects with a global network of indigenous land rights experts and then constructs an effective strategy for implementing its "original" bioregionalism. A member of the Haida Gwaii nation is also a Canadian citizen and contributor to a global indigenous rights community; ecological place is connected to a national and global identity trajectory.

To further our agenda of enhancing environmental learning and make it relevant to the Anthropocene, in the next section I'll describe some educational approaches that help us better understand the relationship between local and global, how we encounter difference, and what our obligations may be to the world beyond ourselves—that is, ways to imagine an ecocosmopolitan identity.

Maps and Stories

Maps become truly interesting when they help to show the audience that which usually remains invisible.
—Antonis Antoniou, Robert Klanten, and Sven Ehmann, *Mind the Map*

Maps have been weapons of imperialism, it is true, but they have also been used effectively to reclaim and defend millions of square miles of indigenous territory.
—Mark Dowie, *The Haida Gwaii Lesson*

In this section, I offer some learning approaches that illuminate many of the concepts covered in this chapter—a more hands-on guide to cosmopolitan bioregionalism. Readers should use these ideas as templates, depending on scope and setting. Surely they are applicable in formal educational settings, but I hope they are also explored in community settings, at planning meetings, and with various types of nongovernmental organizations, including activist and networking groups as well as foundations and philanthropies. I encourage you to try some of these ideas as relevant in your own life too, as prompts for writing, mapmaking, journal keeping, public art, interviews, memoirs, or any kind of artistic or community-based endeavor. These approaches can be flexibly mixed with some of the ideas in the "Mapping Migration" and "Learning to Navigate Networks" sections in previous chapters. Moreover, they work best when integrating ecological, cultural, and psychological perspectives. The spirit of these activities is to promote collaborative efforts to illuminate cosmopolitan bioregionalism, a concept that becomes more vital as we better understand and apply it. The activities are divided into a seven-step conceptual sequence.

If these are projects you contemplate trying, I highly recommend that you visit the website of some exemplary visual artists and cartographers to get a sense of what's possible, and the amazing ways of coordinating the observational and interpretative activities listed below. Here is a brief list of what is a global catalog of such artwork and graphic design. You can either visit their associated websites or place their names in Google Images to get samples of their work.

See Phillipe Rekacewicz's maps and others at LeMonde Diplomatique (https://mondediplo.com/maps). The magnificent series of atlases by Solnit and her cocreators cover many of the themes in this book. Check out the maps of Brooklyn Villages and Mother Tongues in Queens in Solnit and Jelly-Schapiro's *Nonstop Metropolis: A New York City Atlas*. Australian artist Alex Hotchin has a remarkable portfolio of sketch/map hybrid displays (http://alexhotchin.com). Dutch artist Jan Rothuizen also has a unique sketching and mapmaking storytelling style that highlights urban diversity (https://janrothuizen.nl). Olalwkan Jeyifous, a Nigerian-born, Brooklyn-based artists, uses narrative maps to illustrate urban planning strategies (https://archinect.com/vigilism). Jennifer Marravillus offers an incredible display of what she calls memory maps, an essential aspect of her cartography practice (http://jenmaravillas.com). And take a look at the cartographer Molly Roy's excellent portfolio, especially her maps for the Solnit series (https://www.mroycartography.com).

Now on to the activities.

1. What Is Indigenous?
Indigenous has an interesting etymology, derived from the Latin "indiga," meaning sprung from the land, or nature, and from "gene," or to give birth or beget. Who did this land give birth to? That's an interesting question

to start with, for sure! Depending on where you live, it's instructive and often inspiring to refer to the creation story of the most recent aboriginal culture, and then review its ecological practices. What resources did that culture use? What did these people eat? How were their settlements constructed? How did they travel, and where did they go? What were the original place-names? Where are they now?

A second phase is to trace the various communities that have moved in and out of the place since the aboriginal habitation. Who came, who left, and who stayed? For how long were they residents? How did they use resources? What were their claims on the land? What conflicts and agreements emerged as different groups laid claim to the place?

A great activity is to create a map that superimposes original place-names with contemporary ones.

2. Travelers and Newcomers

What is a newcomer? Are there criteria for welcoming newcomers? A newcomer may be a traveler (someone who is just passing through), new resident (someone who just moved to the community), outsider (someone who is only peripherally engaged in a community), or migrant. What about affluent people who own multiple homes, including one in your community? What is their status? How is it different from migrant workers who follow the trail of seasonal employment? What about resident, migratory, and/or invasive species? Is there a correlation between human movement and the lives of species? Organizationally, a newcomer may be a new business or institution. What are the circumstances of their arrival and departure? By what criteria are they allowed entry?

Good travel literature involves narratives of people who are just passing through, yet it provides compelling first impressions. What do you observe when you are a traveler? How are you treated when you enter a new community? How does that compare to how you treat newcomers when you are at home?

3. Mapping Diversity

Another version of this activity is presented as migration in your community in the "Mapping Migration" section. The difference here is that the focus is on the multiple forms of diversity, not just ethnic background.

What is diversity? The etymology is fascinating. The Old French (*diversite*) refers to uniqueness, distinction, variety, and oddness. The Latin (*diversitatum*) refers to contradiction and disagreement. Diversity entails difference and surely can lead to misunderstanding. The challenge is to first define diversity and then find it. What are the criteria by which you assess diversity: clothing, food, skin color, language, sexual preference, and/or political view? Are you intrigued by diversity or fearful of it? How do you encourage or avoid it in your daily routine?

4. Where Are the Contact Zones?
Again, we start by interpreting the expression. What is contact? How does it happen, and where does it occur? What distinguishes a contact zone? When do you frequent them, and what happens when you do? What's the scale of a contact zone? Is it an entire city, place where people meet, school, business, or community center? Do you seek out contact zones or avoid them? Are they safe places?

5. Encountering Difference
As a follow-up to the two activities above, what happens when you encounter difference? Are there particular circumstances that enhance learning in situations where difference abounds? What restricts learning? What catalyzes understanding or misunderstanding? Are there educational and community-oriented best practices for encountering difference?

6. Connecting Local and Global
The essence of cosmopolitan bioregionalism is a deep understanding of how local and global intertwine, and how to enhance local, place-based decisions while accounting for global relationships. What is distinctive about a place, and how is it enhanced with a cosmopolitan perspective? Here are three avenues of exploration.

Where does food come from? Do an assessment of locally available commodities by comparing what's available in a local food market (let's say a coop) and supermarket. It's fascinating to map the aisles of a market and trace the original locations of the food on the aisles, or in some cases, the many locations that are necessary to produce an item of food.

Where does media information come from? What news sources are followed in a community? You can organize this by asking a group of people to list the various places they get information from and then trace the origins of that information, both in terms of the type of media and location of the actual source of the information. Or if you are working with a group, ask each person to interview ten people to get these data. With even a small group you can assess over a hundred sources and quickly gain an interesting geography of information. The interpretation can focus on the extent to which the information is generally local, regional, national, or international, how to trust the validity of the source, and whether the information is biased toward a local or global view.

What is your identity trajectory? Consider the five identity trajectories I covered in the "Contact Zones" section and determine the extent to which they influence you. This can be done with a chart or map by comparing your place-based affiliations. This is particularly interesting with a group of people as it triggers excellent conversations about local and global identity configurations, and often breaks down barriers of exclusion.

7. Ecocosmopolitan Sense of Place

As a culminating activity, I suggest an artistic ecocosmopolitan sense-of-place project as a way to think about both local place and its influences, and then the various ways that cosmopolitan characteristics enrich that place. One way to do this is to invite participants to create a patch that signifies what's important to them about the place or places where they lived, and then to combine all of those patches into a tapestry or montage. Some years ago I visited a museum in Hobart, Tasmania, and explored such an exhibit. The curators asked people who were originally from Hobart but now living internationally to create a small patch (12 inches by 12 inches), and then mail the patch to the museum. The curators then created a beautiful tapestry—a Tasmania-based collaborative sense-of-place map, compiled from global residencies. There are many ways to organize such a project, but this is a compelling example. Consider improvising your own approach.

IV
To Know the World

8
Improvisational Excellence

Improvisation and Environmental Learning

If we jam with the world with the same intelligence and awareness as a skilled jazz musician, we stand a chance of learning a way into the great improvised complexity of the natural world, a concert so immense and endless that we may never be humble enough to accept our role in it.
—David Rothenberg, *Sudden Music*

One of the classic (and still-pertinent) global environmental change texts, *Global Change and the Earth System*, presents a scientific challenge to its readers: "How can an innovative and integrative Earth system science be built?" In a comprehensive final chapter, the authors describe the conceptual and educational foundations of this challenge. Several themes are relevant here. First, they suggest that "the Earth is currently operating in a no-analog state." What they mean is that the magnitude and rate of global environmental change is unprecedented, or "outside the range of the natural variability exhibited over at least the last half million years."[1] If we apply this to the main theme of this book, the future of environmental learning, we can say that we are entering a great unknown. Second, the authors argue that a no-analog state requires that we learn how to cope with complexity and irregularity, stressing that concepts such as nonlinearity, thresholds, uncertainty, emergent properties, indeterminacy, and scaling should be the educational foundations for global environmental change research. These are challenging concepts, and I will briefly explain them later in the chapter. Under conditions of inevitable change and disruption, we need to constantly reassess what we know, how we learn, and how we think.

Throughout this book, I've been exploring many dimensions of unfolding uncertainty—the ramifications of the tides of change. I've addressed how these changes impact both personal lives and environmental learning, including the powerful attraction of the internet and social media, proliferation of networks, movement of people and species around the planet, and challenges of cosmopolitanism, all of which are manifestations of global environmental change. I've emphasized how environmental learning provides an approach to these issues that grounds experience through observation of the natural world, stimulates a broadening perspective, allows us to celebrate life, and generates meaning in the face of unprecedented change. These last two chapters are a culmination of this unfolding thought process, presenting a philosophy of environmental learning predicated on improvisation and reciprocity.

Why are these approaches so important? Improvisation is the essence of adaptive learning, sparking the imagination, stimulating creativity, flow, innovation, and simulation, supplying the inspiration for ideas and solutions, emerging from spontaneity, play, vision, and the unconscious. Improvisation encourages learners to create, experiment, converse, apprehend patterns, encourage collaboration, and most important, adapt to ever-changing boundaries and circumstances. Yet improvisation also relies on structures (knowledge) and skills (practice) that ground learning in tradition and craft.

Reciprocity is the essence of generosity—the recognition that the biosphere is a complex assemblage of interacting organisms that respond to the changing circumstances of ecology and evolution. Every living thing on this planet perceives its environment, and the more we understand multiple perceptual realities, the better equipped we are to understand human flourishing, recognize species interdependence, and open our minds to the possibilities of environmental learning. I will address reciprocity in the next chapter.

Here I'll focus on improvisation. How do we contemplate the future of environmental learning, extraordinary uncertainties of planetary change, and magnificent challenge of human flourishing in the ever-changing biosphere? Shall we agree, for starters, to acknowledge the cloud of unknowing and impossibility of fully understanding the conditions of our existence, so as to strip ourselves of certitude and find the moral courage to celebrate humility in a daunting universe? May we simultaneously celebrate science, the pursuit of knowledge, and the human ability to observe, interpret, and analyze patterns, so as to develop theories, assert propositions and hypotheses, and respect the great capacity of human beings to learn about the world? Shall we agree that the world's great knowledge systems—from science, to various religions, to moral beliefs and values—all share insights, ultimately derived from human experience in the biosphere? Can we open our minds enough to engage in deep conversations about different ways of knowing while celebrating the creative imagination, without giving up what we hold to be dear and true? If we can agree to these terms, then we are ready to engage in improvisational learning. How do we strive for improvisational excellence?

I'll open the conversation with an improvisational walk through the woods. I often start with a plan, but what matters more is my ability to stray from that plan. I'll use that experience to explain what I mean by improvisation and why it's crucial for environmental learning. In the following sections, I'll discuss the relationship between structure and improvisation, while highlighting some of the structures of knowledge that are fundamental for observing environmental change. I'll do so with a sequence of five questions. How might we use language as an educational tool for improvisational environmental learning? What's the relationship between observation and improvisation? How is environmental change science developing an exciting new epistemology of pattern, requiring a purposeful balance between improvisation and structure? How do random processes help us deepen our understanding of pattern-based learning? What's the relationship between improvisation and adaptation?

Our conversation will then focus on the flow element in improvisation and state of immersion that results. I'll briefly touch on the prospects for networked improvisation and how that contributes to environmental change research, and how that is made possible with the open-minded, collaborative conversations that are crucial to civic discourse in a democracy. A good conversation leaves room for improvisation.

The chapter will conclude with a discussion of ecologically oriented improvisational knowledge systems, what they might look like, and how we might construct them. And I'll open that conversation by reviewing the *I Ching*, a timeless wisdom guide that emerges from symbolic biosphere archetypes.

An Improvisational Life

Recall how the Dao De
Jing puts it: the trail's not the way.
No path will get you there, we're off the trail,
You and I, and we chose it! Our trips out of doors
Through the years have been practice
For this ramble together,

Deep in the mountains
Side by side,
Over rocks, through the trees.
—Gary Snyder, *No Nature*

I believe there is a subtle magnetism in Nature, which, if we unconsciously yield to it, will direct us aright. It is not indifferent to us which way we walk.
—Henry David Thoreau, "Walking"

I recently took a walk in the woods on a rainy, cool, late September morning, enjoying the soft, moist breeze. At this time of year, the Monadnock region, the hill country of southwestern New Hampshire, is in the early stages of what will become a rapid and intensely colorful foliage display. Often when I take a walk I have a plan or tentative destination. But if something unexpected beckons or there is an intriguing fork in the road, I may decide to change my course.

As I began my walk, I just wanted to breathe the air, move my body, and read the day. A few hundred yards from the house there's a protected wetland, originally a beaver pond, nestled like a small bowl against the backdrop of Mount Monadnock. The wetland can't be seen from the road. There's one subtle opening that brings you to the mucky edge. It's easy to walk past it without knowing that it's there, yet I noticed a flash of color. So I decided to check it out. It was a great move. Around wetlands, autumn foliage tends to be advanced, as the trees, especially the maples, are exposed to more environmental stress. The maple trees around the beaver pond were all turning red, a vivid and spectacular display of color against the background of the dark green forest. Patches of yellow and orange appeared too.

I didn't intend to visit the wetland, but I caught a glimpse of the colors. Of all the paths I might have chosen, this is where I responded to the subtle magnetism and traveled off the trail. This simple decision guided me into a new realm of observation. I spent the remainder of my walk skirting the edges of the wetland, noticing the patterns of the leaves falling into the shallow water, observing how the foliage revealed the biogeographic distribution of trees, watching the breeze gently ruffle the patches of open water, and then listening carefully to the sounds of birds, a few lingering wood ducks, migrating hawks, and some resident chickadees. I looked to the sky and saw the first breaks in the clouds, indicators of a cold front that would eventually clear the sky and blow away this misty morning.

While standing on the wetland edge, I thought about the many transitions of the landscape—the boundary between land and water, zone between shrubs and trees, and mushy, mossy ground in between. I thought about transitions in the passage of time—the last green vestiges of summer, the cooling air, and the foreshadowing of the future. Soon the entire forest would be aglow with the reds, yellows, and oranges of the spectral autumn landscape. Openings in the cloud cover portended a night of clearing and the possibility of tomorrow's sunshine. I shifted my attention and contemplated the longer passages of time. This wetland was formed by beavers who built a dam at the opposite end from where I was standing, flooding the trees and creating environmental stresses while forming a new landscape.[2] This landscape was deforested in the nineteenth century, when the hill country of New Hampshire was used for sheep farming. Old stone walls mark the property boundaries. Far longer ago, it was covered by a glacier whose remnants are evident in the many large boulders. I wonder, too, what this landscape will look like one hundred or one thousand years from now, when it's likely that a new climate regime will dramatically change the species and plant distribution. There are already clues to those changes in the various "invasive" species that line the wetland—travelers from other habitats that are making their presence felt here. I was "riffing" with the wetland, using its prompts to explore the present moment and its farther reaches.

I am describing an improvisational thought process, linked to the observation and interpretation of environmental change. I am revealing what I consider to be the elements of an improvisational life: a way of thinking that delights in having a plan and structure, and then using it to travel off the trail. This process is crucial to my educational development, not only for environmental learning, but for all my most passionate interests and experiences—as a musician, player of sports and games, hiker and bicyclist, and writer—and my social relationships in organizational life. In this chapter, I will focus on improvisation and environmental learning, but I wish to emphasize that all these learning spaces and interests contribute to as well as inform each other, and are crucial to, as Mary Catherine Bateson once put it, "composing a life."[3]

So what is improvisation? John Ayto's *Dictionary of Word Origins* suggests that "etymologically, if you improvise something, it is because it has not been provided for in advance." The word emerges from the French *improviser*, descended from the Italian extempore, and then further descended from the Latin *improvisus*, meaning "unforeseen."

Amazingly, Ayto reports that the first recorded use of the verb in English is in a Benjamin Disraeli novel from 1826.[4]

Here's my definition of improvisation (there are many of them out there). Remember that the point of all this is to enhance and unfold how improvisation contributes to environmental learning. Improvisation is the ability to spontaneously respond to dynamic changes in the environment by adapting structures of knowledge to new contingencies while playing with forms, patterns, and ideas as they spontaneously emerge.

When observing nature, I'd rather do so by looking at the everchanging patterns, such as clouds floating across the sky, leaves rustling in the wind, water flowing down a stream, ocean waves crashing against the shore, migrating birds moving through the trees—biosphere phenomena that emerge from structural relationships, but are always different, an infinite variety of possibilities and pathways, changing according to the prevailing conditions. This is the essence of environmental change. Repeating patterns and structures that are constant enough to be noticed, yet are never quite the same. I am less patient with the painstaking but equally important observational skills of identification and order. Yet this is a crucial aspect of environmental learning, and without it I would be floating in a world of dynamic change with no ballast or anchor.

Both Snyder and Thoreau, whose lovely passages frame this section, are and were, respectively, excellent naturalists, constantly identifying species as well as determined to grow their ecological and natural history knowledge through study, analysis, and a respect for scientific inquiry. Environmental learning is a balance between structure and improvisation.

Language and Place

In his magnificent book *Landmarks*, Robert Macfarlane cites his concern that in a recent edition of the *Oxford Junior Dictionary* many "nature" words were culled and replaced by "tech" words. For example, acorn, heron, and pasture (among others) were replaced by blog, broadband, and chat room. Macfarlane writes that "the substitutions made in the dictionary—the outdoor and the natural being displaced by the indoor and the virtual—are a small but significant symptom of the simulated life we increasingly live."[5]

This inspired Macfarlane to develop and curate a comprehensive glossary. He compiled "thousands of words from dozens of languages and dialects for specific aspects of landscape, nature, and weather."[6] These words are mainly specific to the British Isles along with the great variety of landscapes, languages, and cultures that contribute to its heritage. There are nine glossaries—flatlands, uplands, waterlines, coastlands, underlands, northlands, edge lands, earth lands, and woodlands. And each of those glossaries is subdivided accordingly. Here are the subsections for the coastlands glossary: bays, channels, and inlets; cliffs, headlands, and defenses; currents, waves, and tides; fishing and boats; lights, hazes, mists, and fogs; the sea; and shores and strands.

Let's swim a little deeper to explore a few of the words in the currents, waves, and tides section. *Adnasjur* (Shetland Islands) refers to a large wave or waves, coming after a succession of lesser ones. *Sruthbladh* (Gaelic) is the violent motion of waves advancing on and receding from the shore. Here are a few words from the lights, hazes, mists, and fogs section. Woor (Manx) is the low-hanging sea mist that dims the light and chills the air. Glimro (Orkney) is a phosphorescent glimmer.

These words, like most in these glossaries, are place-based, grounded on observations that build on generations of perceptual experience. As you flip through the pages of *Landmarks*, you encounter patterns that you've observed but never thought to name. And you encounter descriptions of phenomena that you've never seen because they occur in an unfamiliar landscape. I love flipping through the glossaries because by doing so, I sharpen my perceptual skills, and feel empowered to observe patterns that I didn't think had any real significance or I hadn't ever noticed.

Consider that Macfarlane's curated lexicon is specific to the British Isles, and although it may be relevant to similar landscapes in other realms, it doesn't cover many of the world's landscapes—deserts, savannas, ocean floors, volcanoes, jungles, and so on. You could create similar glossaries for just about any place on earth, resurrect the treasure of indigenous languages that describe those places, and develop a most spectacular knowledge base of landscapes and environmental change. Sadly, many of these words are disappearing and will soon be lost to human awareness.

These place-based words provide a wonderful catalog of observations, patterns, and interpretations of environmental change. You start with what you closely observe that's right in front of you. Anyone who has spent time gazing at the sea has witnessed adnasjur. Indeed, the more closely you observe patterns of waves as they break on the shore, the more likely you are to find patterns within those patterns, eventually being able to anticipate what may come next. As you develop perceptual depth and observational awareness, you develop a lexicon of patterns that help you assess, interpret, and even anticipate environmental change.

Inevitably there will be anomalies, and when they occur, you will be able to recognize them. The most extraordinary example of an adnasjur is a tsunami. Will familiarity with an adnasjur enable you to anticipate a tsunami? The Chao Lay, the so-called indigenous sea gypsies of Thailand, reportedly anticipated the catastrophic 2004 tsunami because they were so familiar with the patterns of ocean waves.[7]

I believe the key to understanding environmental change is to develop the capacity to detect, observe, and analyze patterns, and then apply that knowledge as widely as possible. This is what I mean by pattern-based environmental learning. As I'll explain shortly, it's also the foundation for improvisational excellence.

I appreciate MacFarlane's approach because it's so accessible. These wonderful glossaries open the doors of perception. Macfarlane notes that in the process of experiencing his project, "life and language collapsed curiously into one another." He discloses that many of the subjects of his book (both the glossaries and connecting essays, based on his own explorations and the words of literary guides) "are often best represented not by proposition but by pattern, such that unexpected constellations of relation light up."[8]

Macfarlane's *Landmarks* has hopeful and inspiring educational application. "We may have forgotten 10,000 words for our landscapes, but we will make 10,000 more given time. . . . And this is why the penultimate glossary of the book is left blank, for you to fill in—there to hold the place-words that have yet to be coined."[9]

In *Language Making Nature*, David Lukas does exactly what Macfarlane suggests. He proposes a series of alternative field guides, or approaches to environmental perception that allow people to bathe in the direct experience of species and landscapes. Lukas challenges readers to create their own words to express how they perceive the biosphere. He provides various ways to deconstruct, re-create, and reorganize language by breaking words down and then putting them back together again. He explores root forms, affixes, suffixes, word origins, the use of vowels and consonants, and the use of older forms of words as taken from Old English, Anglo-Norman, French, Greek, and Latin. Lukas provides an outline of English word formation, utilizing affixation, compounding, conversion, blending, clipping, and word manufacture. And like any good field guide, there are many lists and examples to identify and then examine.

Language Making Nature moves beyond being a set of instructions to an educational inspiration. There are sections that propose topics for word creation with some helpful suggestions for getting started. Short chapters on place-name elements, texture, rock, seeing color, states of mind, wind, clouds, and species elaborate the possibilities. The essence of his project is neatly summarized in the section "Four Ways to Create Original Words": "Borrow or repurpose old words or change them into new words. Add prefixes or suffixes to preexisting words. Convert a word to a new part of speech. Combine two or more words into a new compound word."

And then Lukas adds a dose of philosophical instruction.

Bring every word to life:

—Know its roots
—Place it carefully
—Have it mean something.[10]

On a brisk October day in the midst of foliage season in the New Hampshire woods, I wondered if I could utilize the Lukas method to expand my perceptual awareness through improvisation—the essence of pattern-based environmental learning. I was sitting on my porch when a dragonfly landed on my shirt. It flew away and then returned, landing on the white pages of the Lukas book. I had several moments to observe it. After its departure, I found my *Stokes Beginner's Guide to Dragonflies* and identified it as a band-winged meadowhawk.[11] But then I thought, Have I learned nothing? Give it your own name. And I realized what a poor observer I am. I would have to know much more about its life habits and habitats before I could give the dragonfly a fitting name. Maybe, I thought, I can give it many different names until one finally fits. Maybe I am caged by the language rules that I am familiar with, and as Lukas contends, the only way to break away from the blinders of rule-fixed language is to develop my skills. So I temporarily named it a Monadnock meadowhawk, linking it to my place.

Although I didn't feel ready to use words to describe new impressions, I thought I might get partway there by writing down impressions of phenomena that I had no single word for. So here's my list of foliage observations deserving of creative words or thought forms:

- the pointillist granularity of countless spectral foliage arrangements
- how the wind and shadows cause movement and motion while deepening the vibrational color forms
- the changing hues of blue sky through the canopy shadows
- multiple colors of leaves on the same tree, and the light of the sun reflecting differently off variably colored leaves

- passing clouds over the bright canopy
- colorful leaves falling like floating petals of rain
- how the forest floor becomes a more complex mosaic with each falling leaf
- a sequence of foliage arranged in the order of the spectrum
- the sound of insects on the last warm October day

I may not have invented new words, but the spirit of the Lukas inquiry challenged me to observe more deeply, explore patterns, and allow myself to use improvisation to inspire both processes. Improvisational learning occurred when I entered the observational flow of the moment and used patterns I recognized to help uncover new ones. The creative melding of language, place, and perception facilitates pattern recognition. Using language as an improvisational tool further enhances that prospect. There's no reason to confine this observational method to linguistic forms. Why not use icons, visual symbols, dance movements, graphic designs, musical notes, musical scale systems, drumbeats, or other hybrid forms of communicating about and playing with our observations of the biosphere?

In the final chapter of *Landmarks*, Macfarlane suggests that a pertinent follow-up project would be a glossary of newly invented words that describe the environmental processes of the Anthropocene: "new-minted Anthropocene terms." He asserts that "such a glossary would need to detail the topographies of toxicity and dereliction that we have made, the spectacles of pollution, corruption, and extinction we have induced, and the miracles of geo-engineering we have wrought."[12]

Are there words for the expanding gyre of plastic bottles in the Pacific Ocean, rapid bleaching of coral reefs, or unprecedented temperate climate microbursts that accompany climate change? Perhaps naming these anomalies allows us to take them more seriously than we do. This is a dark task indeed, but also a necessity for environmental learning. How do we identify and name unprecedented patterns of desecration? And to accompany these terms, can we also develop words and expressions to describe new forms of environmental activism and empowerment, or creative and sustainable solutions? Let's add these challenges to the growing list of important tasks for the engaged environmental educator and concerned citizen.

Improvisation and Observation

"Change is the only constant" is an adage well suited to the landscape of central New England. Whether wrought by human hands, storms, or longer-term

alterations in climate, change has forged a dynamic landscape. As human observers who measure change in days and years, it is difficult for us to grasp that the landscapes we know in our lifetimes are not only ephemeral, but also radically different from those that preceded them. In ways both predictable and unimaginable, the future landscape of central New England will differ from the one we know.
—Tom Wessels, *Reading the Forested Landscape*

I had the great privilege of spending several decades working with Tom, a remarkable community ecologist. Taking a walk in a "forested landscape" with Tom always inspires a blend of observation, interpretation, and improvisation. Just by looking at forest composition, landscape forms, human artifacts, and other environmental variables, Tom develops compelling and scientifically consistent interpretative narratives. He engages his students by asking challenging questions that require focused observations. He demonstrates how those observations yield interpretative insights. In a class setting, Tom's questions build a uniquely collaborative problem-solving state of mind. All the participants feel as if they are contributing to the collective interpretation.

Several years ago, as a complement to a public celebration of Tom's teaching and writing, I conducted a long interview, attempting to better understand Tom's learning process. I consider Tom an exemplar of pattern-based environmental learning.[13]

As the early age of five, Tom wandered the woods outside his Connecticut home. He spent countless hours exploring the local woodland. He learned to navigate the forest on his own. He did so by recognizing various forest types so that he would never get lost.

Amazingly, he never learned how to identify what he was observing, at least until he entered college at the University of New Hampshire. "I do remember when I first learned about field guides. That was so exciting. I remember I got this canvas army satchel that would just fit four Peterson Field Guides. And I'd be out every day with those things or my binoculars. When I went to Colorado (for graduate school), my interest in just ID-ing species waned. That's also when I began to more deeply look at larger landscape patterns." At the University of Colorado, Tom had several mentors who emphasized pattern recognition at the landscape scale, enabling him to develop community ecology skills.

I asked Tom how he assesses a landscape that is brand new to him, where he is less likely to be familiar with species composition and local geology. He explained that he searches for patterns by differentiating between compositional and structural approaches. The compositional patterns help him assess the population of a habitat. The structural patterns help him understand the relationship between species and landscape.

Tom stressed that improvisation is intrinsic to his learning process because then he doesn't have a set script that he's following or something specific that he's expecting. It's important to the observational process because he goes out to a place and is not anticipating anything. "I'm just going out and looking, and then what I see is going to start affecting how I'm looking and where I'm looking. It's constantly being readjusted. So I think that is pretty much improvisational."

He stressed, however, that his improvisational approach always refers to conceptual structures, both his experience and knowledge as a community ecologist, and the methodology of his observational process. Without that structure he wouldn't have any kind of road map. "Improvisation is having flexibility within a structure."

Pattern-Based Environmental Change

Tom's approach to "reading the forested landscape" is an excellent example of pattern-based environmental learning—an approach to observing and interpreting environmental change emphasizing pattern recognition across multiple scales. When we study global environmental change, the only way to make sense of its complexity is to recognize the patterns of change, how they link atmospheric, oceanic, terrestrial, and organismic phenomena, and how they reflect spatial and temporal variation. These patterns are the structures of knowledge that serve as the foundation for environmental change science.

For example, the field of landscape ecology describes ecological spatial variation with an illustrative pattern language (mosaics, gaps, boundaries, corridors, patches, edges, and fragments). Recall our discussion of constructive connectivity and consider those topological visualizations as additional layers—networks, nodes, and links. Consider the patterns of oceanic and atmospheric circulation as another layer of flows and fluctuations. In their outstanding ecology text *Toward a Unified Ecology*, Allen and Hoekstra examine all these patterns at multiple scales—landscape, ecosystem, community, organism, population, and biome.

In observing environmental change, pattern concepts such as waves, thresholds, and cycles are similarly illustrative. Waves, for instance, appear ubiquitously as visual and acoustic representations of rates of change, reflecting frequency, longevity, and periodicity. Waves are tangible indicators of environmental change, as Macfarlane reminded us with his lexicon of wave observations from relic languages (words such as adnasjur and sruthbladh).

The ability to observe wave patterns helps the observer recognize thresholds: the events or flows that reflect a dramatic shift in condition. How do we recognize when a threshold is reached? Is it an indicator of an irreversible condition or merely a discontinuity in a periodic cycle? This understanding is immeasurably helpful in assessing the impact of climate change. Gavin Pretor-Pinney's wonderful book *The Wave Watcher's Companion* is an essayistic field guide that helps readers sharpen their ability to observe and connect wave processes in multiple settings. He poses and then answers exactly the kinds of questions that help us learn to observe more closely: "How do waves form in a storm? And how, for that matter, do they change from that choppy confusion into the ordered procession of crests that tumble up the beach towards you? For the answers, you need to chart their journey across the ocean, to follow each stage of their development, from birth at sea to foaming death on the shore."[14]

Or flip through Tristan Golley's book *How to Read Water: Clues and Patterns from Puddles to the Sea*, another essayistic field guide that explains how careful observation of water helps you understand ecology, geomorphology, weather, and oceanography. Golley opens by noting that "we can look at the same stretch of water every day for a year and not see the same thing twice. How does one compound behave with such diversity? And what do the differences we see from day to day and from one place to the next actually signify? This is a book about the physical clues, signs and patterns to look for in water, whether you are standing by a puddle or gazing out across miles of ocean."[15]

Another intricate pattern language is revealed by studying cycles. When you observe a cycle, such as the carbon cycle, you move through land, water, and sky, countless organisms and habitats, and photosynthesizing plants. Or you may get buried in the depths of the ocean floor and then reemerge millions of years later as a fossil fuel. In CO_2 *Rising*, Tyler Volk writes about climate change by elaborating on the specifics of how a carbon atom travels through all these milieus.

A cycle is a continuous and predictable series of relationships within a system, in which the flow and exchange of materials moves according to repeatable yet highly variable patterns. Of particular interest is the relationship between cycles, which may form another system of cycles or have nonlinear emergent properties. As Allen and Hoekstra write in

Toward a Unified Ecology, "Remember that populations, communities, ecosystems, and landscapes can be considered as the embodiment of cycles." They describe how trees "play" in many cycles at once, and that "structures are the places where the various cycles of nature kiss."[16] Schoolchildren can observe cycles, and yet it is the depth and complexity of cycles that reveals the intricacies of environmental change.

There are dozens of exceptional popular and technical books that emphasize how observing patterns in nature helps reveal underlying structures of environmental change. Among the best is Philip Ball's beautifully illustrated *Patterns in Nature*, depicting patterns that transcend scale and what they teach us about environmental change—symmetry, fractals, spirals, flow and chaos, waves and dunes, bubbles and foam, arrays and tilings, cracks, and spots and stripes.[17] Christopher Alexander's classic architecture text *A Pattern Language* and then his magisterial four-volume The Nature of Order construct an elaborate theory of urban planning and sustainable architecture based on his observations of patterns in nature.[18]

We are living in a dynamic intellectual time with emerging epistemologies of environmental change—how we organize, interpret, and observe ecological variables—and many original approaches to pattern recognition. What do these concepts have in common? They all emphasize that observing environmental change requires conceptual flexibility—the ability to discern patterns and structures in nature, and then how your ability to recognize these patterns also enables you to see beyond them. As Wessels reminds us, the patterns themselves always change. The future of environmental learning will reveal new approaches to thinking about environmental change, requiring open-mindedness, connectedness, creativity, and a spirit of discovery, supported and inspired by epistemology and pattern—the structures that allow for improvisational possibilities.

Improvisation and structure reflect an elaborate dance, a dynamic blend of observational awareness, interpretative skill, and theoretical conjecture. In the final sections of this chapter, I'll speculate on an imaginary catalog of improvisation-based, environmental change field guides and how educators may create their own. Now more than ever, we must stimulate learning approaches that encourage creativity and innovation. They're the essence of inspired environmental learning—sparked, explored, and extended through improvisation. Next I'll explore the concept of random processes, as they often create the conditions for improvisational learning.

Random Process

Chance can matter. Accidental meaning is essential for improvisation, because we are never fully in control. When we don't know where the order of the work is coming from, wonderful surprises can result.
—David Rothenberg, *Sudden Music*

It is important to me that my children can distinguish a vulture from a golden eagle by the cant of its wings. It reassures me to know that they can recognize the evening call of robins and the morning calls of doves, that they know from its tracks whether a rabbit is coming or going, that they always know which way is west. I want them to go out into a rational world where order gives them pleasure and comfort, but also an improbable world, wild with sound and extravagant with color, where there is always a chance they will find something rare and very beautiful, something that is not in the book.
—Kathleen Dean Moore, "A Field Guide to Western Birds"

In the Monadnock region of southwestern New Hampshire, we live in something of a snowbelt. We have just enough elevation (twelve hundred feet) and are close enough to the coast (eighty miles) that many winter precipitation events falling as rain just a few miles to the south or east will become snow at our location. There are many microclimatic variations that determine the form of precipitation. Will it be rain, snow, sleet, ice pellets, or a mixture? Sometimes it seems we're getting "oobleck": the sticky, insidious substance from the Doctor Seuss classic *Bartholomew and the Oobleck*.[19]

The track of the storm matters too. If it's a classic "nor'easter," just a few miles makes a huge difference. If the storm tracks off the coast, it will bring in enough cold air that we'll get snow. If it tracks inland, it will bring in enough mild air that we'll get rain. Sometimes the cold and warm fronts get all tangled up (occluded), and we get ice.

Although the weather forecasters are becoming quite good at figuring all this out in advance, you can never be exactly sure about the track of a storm or the microclimatic variations. When we move from weather to climate, and try to assess atmospheric and oceanic flows at more challenging spatial and temporal scales, the combination of variables reaches new levels of complexity. I bring up the daily winter weather as an example because it's a tangible way to contemplate key concepts of environmental change science. To reiterate those concepts from the International Geosphere-Biosphere Programme's classic *Global Change and the Earth System* study, they are uncertainty, nonlinearity, emergent properties, thresholds, and scaling.

Let's briefly explore these concepts and then link them to improvisational learning by discussing the meaning of randomness. Uncertainty essentially refers to an awareness that the complexity of variables leads to unknown outcomes. Nonlinearity means there is no explicit relationship between a variable and its outcome. An emergent property is when a collective process cannot be predicted from the individual components of that process. A threshold is a bifurcation point when a limit is reached, resulting in a cascading series of events that only occur at that point. Scaling is an awareness that various boundaries of space, size, and time may yield different but related observations.[20]

I'll get back to these concepts momentarily. But let's think for a moment about the meaning of chance and random processes. Actually a deep look into these concepts reveals some highly nuanced philosophical distinctions, including discussions of probability and determinism.[21] What appears random may turn out to be part of a pattern we have yet to perceive.

Of late, the word random has gone viral. Indeed, the online urban dictionary suggests that "random" is used to describe anything a person doesn't understand.[22] "That was totally random" is an expression I'm sure you've heard or have used. It makes sense that people would use random to explain processes that are out of sight or mind.

Improvisational learners embrace random processes for four important reasons. First, they understand that by investigating them, they will discover new patterns of interest. It's snowing today. It's a thick, crunchy, sticky snow, and the trees are covered by an elegant white frame that highlights or illuminates the patterns of the branches in the forest. I can clearly discern the connective tissue (the network) of relationships between the canopy and ground. I can also determine which trees are most heavily weighted by the snow and which are most likely to fall. There is a threshold at which point these network connections become visible. Had the weather system been slightly different, I would not observe these patterns.

Second, a randomizing process can serve to orient your thinking when you are faced with too many variables. In a few moments, I'll go outdoors and have many choices as to where to walk. Any direction will yield interesting observations and insights. How will I decide where to go? I can follow the path (it's much easier that way) and it will take me along my customary route. But what if I decide instead to find a random compass direction and take off according to whatever direction I generate? Sometimes when I'm playing my synthesizer, which has access to thousands of sounds, I'll randomly generate a sound and see where it leads me.

Third, randomness can be something that you work with, challenging you to adapt to changing circumstances. If it had rained last night, the day would look very, very different. But it snowed and I had to change my plans. Until the driveway is plowed, I can't drive anywhere. So rather than driving to town, I'll take a walk in the snow. Possibly later today the weight of the snow will cause a tree to fall on a power line and we'll lose power. Possibly a tree will fall and not cause any damage. It may seem random, but because I live in the forest, the possibility of a tree falling on a power line is much higher than if I lived in town. In fact, the risk of power outage is mitigated if landowners and the electric company take the time to cut down vulnerable trees.

Fourth, randomizing processes (under the right circumstances) are a great deal of fun. They throw your world into temporary disorder and provide you with the opportunity to make the best of the circumstances, whether they seem to represent favorable or unfavorable fortune. Games of chance have a long, long history in human culture. Throw the dice and see what happens. When you combine games of chance with games of strategy, you have all kinds of interesting possibilities for solving problems, building scenarios and simulations, and learning how to think using improvisation. This appeals to the human proclivity for games. A well-designed game allows you to explore the relationship between chance, strategy, and outcome.[23]

In his wonderful book about teaching physics to Tibetan Buddhist monks, *Humble before the Void*, Chris Impey explains his teaching philosophy. "In the spectrum of teaching methods, play is the most neglected. . . . Games that combine constraints with abundant possibilities are extremely useful in conveying the essence of science. The landscape is unfamiliar and liberating when there are no clear expectations and no right answer. After all, we live in a universe governed by simple laws of nature where almost nothing is impossible."[24] Juxtapose that with Gordon M. Burghardt's comments in his classic work *The Genesis of Animal Play*: "In a very real sense, only when we understand the nature of play will we be able to understand how to better shape the destinies of human societies in a mutually dependent world, the nature of our species, and perhaps even the fate of the biosphere itself."[25] In his book about improvisation, Free Play, Stephen Nachmanovitch, finds inspiration through his environmental observations: "Looking out, now, over the ocean, the birds, the vegetation, I see that absolutely everything in nature arises from the power of free play sloshing against the power of limits."[26]

These three passages, taken together, are eloquent reminders that the essence of environmental learning is a blend of observational awareness, interpretative fluidity, and improvisational excellence. You study the patterns and structures of nature only to realize that the "abundant possibilities" that may appear as random processes, or unlimited choices, are a vast playing field from which you can always derive new insights and your creativity will always be challenged.

Adaptation and Improvisation

Adaptation—one of life's emergent properties—exposes patterns, or repeated sequences and correspondences, that are detectable by even the simplest organisms. The patterns that matter are the opportunities and dangers influencing survival and the ability to multiply.
—Geerat J. Vermeij, *The Evolutionary World*

Like Darwinian evolution itself—comprised of chance mutation and natural selection—improvisation often tries or proposes a "solution," and then the environment selects for or against it.
—Stephen T. Asma, *The Evolution of Imagination*

The subtitle of Vermeij's *The Evolutionary World* is revealing: *How Adaptation Explains Everything from Seashells to Civilization*. It's a series of wide-ranging essays, drawing on his extensive experience as a geologist and biologist, with many examples drawn from natural history, especially Vermeij's work with mollusks. Vermeij reflects on what he describes as "the evolutionary way of knowing," suggesting that by better understanding the evolutionary manifestations of adaptation, we can better respond to the challenges of global environmental change.

For example, he writes that "adaptation is the way in which living things make sense of the world—it enables unusual phenomena related to resources, weather, and enemies to be incorporated into a predictive hypothesis, which is retained, transmitted, and refined in subsequent generations."[27] This fundamental hypothesis is resilient enough to offer commentaries on specific policy issues like climate change to philosophical questions regarding order and variability, and easily translated into rules of thumb for environmental learning. Vermeij is asserting that a deeper knowledge of adaptive behaviors, derived from ecological and evolutionary dynamics, is the essence of environmental learning. By understanding these dynamics, we are better positioned to create effective strategies for responding to rapid global environmental change.

Indeed, adaptive management is a significant field of study in its own right, emphasizing collaborative decision making in the face of uncertainty and bringing diverse stakeholders together to construct robust responses to challenging environmental issues. There is a growing research literature, now several decades old, stressing this approach. Resilience studies is another growing field, applying adaptive management theory (and other approaches) to enhance preparations for the inevitable prospects of environmental change.

Adaptive management and resilience studies underscore the policy applications of the broader evolutionary concept of adaptation. And if you reflect on Vermeij's approach as well as the previous discussion of improvisation, it's clear that the ability to improvise (in policy settings) is an important aspect of the necessary iterative process leading to innovative policy solutions. Vermeij's approach to adaptation provides a rich substrate for how we think about environmental learning, and I'm intrigued with how biosphere processes at multiple scales provide templates for this learning process. A more thorough investigation of these possibilities is crucial. Let's encourage such experimentation and investigation in the spirit of advocating for improvisational environmental learning.

There is also a growing scholarly interest in the evolutionary origins of improvisation, imagination, and aesthetics, as these qualities seem uniquely human, crucial to social learning, and at the core of the quest to better understand the origins of human consciousness.[28] I think it's important to refer to this intellectual path as well. Understanding the origins of creativity, innovation, and hence improvisation cultivates a deeper appreciation of what's interesting about human behavior, and inspires educational possibilities. One of the great mysteries of human consciousness is the existence of the imagination, without which improvisation would be impossible.

Despite all the possibilities inherent in these explorations, I'd like to caution against a strictly adaptationist perspective. It relegates the mystery of imagination to a functional explanation. Improvisation undoubtedly confers selective advantage, but there is more to it than that. Some scholars, including evolutionary ecologists, recognize not only the evolutionary necessity of adaptation but also the importance of beauty for its own sake. In *The Evolution of Beauty*, ornithologist Richard O. Prum reminds his readers that much of aesthetic philosophy views the beauties of nature "through an exclusively human *gaze*," and thus we fail to comprehend the "powerful aesthetic agency of many nonhuman animals" He suggests that aesthetic innovation is a coevolutionary phenomenon, whether it characterizes mate selection or cultural transmission: "Once we understand that all art is the result of a coevolutionary historical process between audience and artist—a coevolutionary dance between display and desire, expression and taste—we must expand our definition of what art is and can be."[29] Rothenberg makes similar claims in his remarkable trilogy of books about music and birds, whales, and insects. In the first book in the series, Why Birds Sing, he argues that birds sing simply because they like to. Whales do too.[30]

This notion stimulates an exciting inquiry. Is improvisation a coevolutionary art-making process, a cultural path that has a life of its own, beyond any prospects for selective advantage, but rather a way to sing the praises of the biosphere? I can't answer this question. Yet I suggest the question itself has educational implications and bears directly on our discussion of improvisation. My stance is that improvisation should be viewed from both perspectives: an adaptive strategy that responds to changing environmental variables and creative process inspired by the sheer joy of it. These processes inform each other, and that awareness enhances environmental learning.

From an adaptationist perspective, improvisation is a process by which we learn (at the organismic scale) how to perceive the ecological moment so we can cope with environmental variability. At the community scale, an ecosystem is a collection or perhaps "ensemble" of individual "players" who similarly explore environmental variables. These dynamic relationships, taken together, and perceived as the interactions between organisms, communities, ecosystems, and the biosphere, are both mysterious and wonderful, subject to analysis, and replete with complexities and intricacies. Adaptive management is a community learning process that aspires to assess these variables to enhance both human flourishing and ecosystem health.

Improvisation serves multiple educational functions. It's a way to practice environmental awareness, explore perceptual possibilities, experience the moment, express the sublime, and appreciate life. It's both a teaching tool and form of learning (just for the sake of it). This is why improvisation is the connector that integrates my lifetime passions: observing nature, playing music and sports, and cultivating healthy organizational dynamics. Ask yourself why you improvise, and you may come up with many of these same reasons as well as others that are unique to your experience.

The connective tissue, the inspirational substrate of improvisation, is that it stimulates a so-called flow experience: a state of mind in which you are totally immersed in the moment, responding to change, activating your whole being, body and mind, and participating fully in your surrounding environment—for a feeling of total engagement.

Improvisation as Flow

The river flows.
It flows to the sea.
Wherever that river flows
That's where I want to be.
Flow river flow.
—Roger McGuinn, "The Ballad of Easy Rider"

Some of my most exhilarating life moments occur when I am riding downhill on my bicycle, at a comfortable enough speed that I can balance concentration between my experience of the bike and landscape. In those instances, I feel as if I am merging with the landscape. I am totally immersed in the scene, my senses are awake, and I am very, very happy. Such moments are life highlights. And they are even more satisfying when they take place in group settings. I relish the times when everything was in sync on the basketball court. Team members developed a collective awareness that seemed to come out of nowhere. But it was there. It was palpable. As an individual player, any excellence I may have temporarily achieved emerged out of that flow. The same is true when I've played music with friends, and we've entered a beautiful improvisational space where all the musicians were listening to each other and we became a collective instrument. Where did the notes come from? Who was playing them? How did we know what we would play next?

The word often used to describe this state of mind is "flow," a mysterious awareness when everything you do fits together, time stands still, and you are maximally engaged in whatever you are doing. Mihaly Csikszentmihalyi wrote an influential book, *Flow: The Psychology of Optimal Experience*, describing this phenomenon, and how it leads to both peak performance and happiness. We can all recall times when we've experienced a flow state of mind. And we are frequently drawn to activities that promise this flow state.

I love the word itself. Say it several times. Flow. Flow. Flow. It has an onomatopoeic quality. The river flows to the sea. We are surrounded by flows—streams of water and wind, tributaries of migration and movement, and waves of energy—eddies, currents, circulations, and cycles. Flow is an engaging biosphere metaphor. There's a wonderful book about it, written in 1962 by Theodor Schwenk, *Sensitive Chaos: The Creation of Flowing Forms in Water and Air*. Schwenk portrays the aesthetic dimension of flow, suggesting that there are parallel streams of wisdom, and the ability to observe flow brings that wisdom closer. The expression "go with the flow" developed new currency in the 1960s along with a renewed interest in Taoism, one of the world's great wisdom traditions that offers poetic approaches for meditating on that expression. Some quick research into the origins of the expression reveals that varieties of go with the flow show up in thinkers as diverse as Marcus Aurelius and William Shakespeare.

Jazz musicians, the ultimate musical improvisers, perform best when they arrive at a flow state of mind. Paul Berliner, in his classic enthnomusicological study of jazz improvisation, *Thinking in Jazz*, provides dozens of narrative and musical examples of creative processes that emerge from flow experiences. Physicist Stephon Alexander, in *The Jazz of Physics*, maintains that there is a direct relationship between physics and music, and looks at the process of discovery, the improvisational learning style that inspires his discoveries in both domains. The study of vibrations, response, and harmony is also the study of how matter and energy flow. In *The Evolution of Imagination*, Asma has an evocative section, "Zen, Flow, and Brain Systems," in which he reminds us that "part of being an imaginative improviser is losing your self."[31] I don't want to give the impression of cherry-picking examples here, as each of these studies reflects on the multiple states of mind that contribute to improvisation. It's a complicated and mysterious process. Indeed, there may be many types of flow experiences, and many different ways to get there.

The environmental literature, from classic Chinese Taoist poetry to contemporary expressions, is replete with testimonies about how natural settings can induce flow experiences. David Abram, in his superb essay "Creaturely Migrations on a Breathing Planet," writes about his experiences on the coast of Alaska that led him eventually to observe spawning salmon: "I lost myself in some reverie or other, until my awareness was brought back to the place by a pale glow spreading into the sky from the rocks on the far side of the stream. The glow got steadily more intense until, as I watched, the full moon was hatched from those rocks, huge and round as a ripe peach, pouring its radiance across the stony beach and the gleaming waves and the rustling spruce needles and generally casting a kind of spell over the whole place."

What I'm asserting is that when you go with the flow, you are entering a world much larger than yourself, a collective ensemble, a team, or perhaps an immersion in the streams and currents of the biosphere, or maybe even access to the waves and vibrations in your immediate vicinity that are clues to the depths of an infinite cosmos. Improvisation is a learning tool that helps us achieve that state of mind. And that state of mind generates deep insights about a world outside your head.

All the works cited above explore improvisation as a learning tool that integrates creativity, discovery, awareness, and knowledge. They all stress the importance of structure and practice as preparation for improvisational excellence. It seems contradictory to say that an improvisational flow state requires preparation. But it does. A basketball team would never get into "the zone" if the players came to the game without years of skill development and practice. And jazz musicians would never accomplish creative improvisations without years of practicing their instruments. Improvisational environmental learning, whether in the service of developing adaptive management strategies or cultivating deeper personal awareness, won't happen without knowledge of the biosphere as a living system. All these processes are enhanced in collaborative settings. I'll look next at how such collaboration is crucial to environmental change research and education.

Networked Improvisation

This is a time when the multifarious worlds of music and art are beginning to meet and blend and create whole new species. We are now seeing a renaissance of crossover art of all sorts. East meets West, popular meets classical, improvisation meets tightly scored composition, video meets digital synthesizer meets Pythagorean monochord meets Balinese trance dancer. Whole cultures can play together, contribute to each other, fertilize each other.
—Stephen Nachmanovitch, *Free Play*

Nachmanovitch is describing a potentially creative impact of globalization: the prospect of intercultural blending and fertilization, possibility of sharing musical ideas across electronic networks, and extraordinary access that people now have to diverse musical explorations. There are now consortia that promote networked improvisation, and people who play music together using teleconnections. This is especially popular in the world of electronic musical composition, in which performers and composers collaborate online. Such collaboration is always enhanced with face-to-face encounters, but many of these encounters often begin electronically.

The same process is happening in science. Michael Nielsen contends that we are entering a new era of networked science and the internet is changing how we share data, making it possible to organize, disseminate, and collaborate on scientific challenges that require multiple scholars and diverse disciplines. Nielsen cites dozens of examples across the sciences where such networked collaboration is occurring, including Polymath (a collective math problem-solving process), GenBank (an online repository of genetic information), and the Galaxy Zoo (where astronomers classify galactic images).[32]

He argues that we are "still in the early days of understanding how to amplify collective intelligence," he is convinced that global connectivity has the potential to inspire a new era of scientific discovery. Nielsen is particularly interested in how to design collaborations that accelerate serendipitous encounters. "Ideally we achieve a kind of conversational critical mass, where the collaboration becomes self-stimulating, and we get the mutual benefit of serendipitous connection over and over again. It's that transition that is enabled by designed serendipity, and which is why the experience of designed serendipity feels so different from ordinary collaboration."[33] Although Neilsen doesn't use the word improvisation, it seems to me that his intention in advocating for designed serendipity is to support learning environments that cultivate improvisational learning.

How can networked science promote pattern-based environmental learning? And what is the role of improvisation in that process? The National Ecological Observatory Network (NEON), founded in 2011, is a National Science Foundation–funded initiative that includes eighty-one field sites across the continental United States. Its primary purpose is to provide continental-scale environmental data, develop an infrastructure for environmental change research, and craft educational tools to work with large data sets. The network describes its research process as a "bold effort to understand and forecast continental-scale environmental change" while engaging the "next generation of scientists."[34]

NEON emerged from the Long-Term Environmental Research Network (LTER), founded in 1981.[35] The LTER was originally conceived as a place-based approach to assessing long-term environmental change, with the intention of broadening an understanding of spatial and temporal patterns. The network branches into the humanities as well, specifically with the Ecological Reflections program that brings humanities scholars to the various research sites. The LTER organizes an "all-scientists" conference every three years. One of its major objectives is to coordinate collaboration among networks, and hence it serves as a network of networks for researching and educating about environmental change. It explains its conference objectives this way: "As long-term researchers, we value persistent attention to what often goes unseen and unrecognized: slow cycles, legacies, the role of rare events. We steward our data and methods with care, because only a community of researchers has a chance of revealing the complex whole. Our work is a marathon, not a sprint, and we take this time to appreciate our fellow marathoners."[36] The conference features an "Idea Cafe." Participants are encouraged to offer three-minute presentations about emerging ideas or projects that may not be fully formed. And they are urged to present them in whatever creative milieu makes the most sense.

In my view, the development of NEON, LTER, and similar consortia, both in North America and globally, is the environmental change research equivalent of Nachmanovitch's observations about the sharing of world music. It's a recognition among scientists, humanists, and educators that to enhance, stimulate, and cultivate environmental learning, we need multiple minds and perspectives to better understand the patterns of environmental change. And that learning process requires creative collaboration, improvisation, and a willingness to connect with colleagues around the globe. These collaborations, in turn, require open conversations, good listening, and a willingness to share information in the service of a greater good. Constructive dialogues can promote improvisational learning and are fundamental to civic discourse in a democracy.

Improvisation, Conversation, and Democracy

The improvising imagination is one of the greatest tools of the open society. It cannot guarantee democracy or liberal values, and it cannot by itself resist the schemes of autocrats and tyrants, but left to its own nature, it champions difference and experience.
—Stephen T. Asma, *The Evolution of Imagination*

Only with the possibility for endless improvisation does the world's heavy history leave room for us to move forward and change it.
—David Rothenberg, *Sudden Music*

Think about the virtuous qualities of a great conversation. When people are listening carefully, genuinely aspiring to collective insights, and feeling the value of their own contributions, then a vibrant discussion creates emergent possibilities. A great conversation generates creative insights beyond the capacity of any individual to develop on their own. The best conversations I've had, whether in structured teaching environments, conferences, or serendipitous circumstances, resemble a flow state. People are free to explore new ideas, feel supported in their contributions, and take pleasure in the creativity (or wisdom) of the group. You no longer identify with an idea as your own. You take delight in the collective insights of collaborative minds.

As we all know, such discussions are not nearly as common as we'd like them to be. Too often, people are mainly interested in their own ideas, take up too much of the conversational space, or are more interested in being right and getting their own point across. Sometimes people who have the most to contribute are hesitant to participate because they don't want to get into competition. One of the most important attributes of a good teacher is to facilitate great conversations. This takes enormous skill and self-awareness. I know that in both teaching situations and other settings, especially when I serve as a facilitator, I have to make sure that I temper my own enthusiasm for an idea to ensure that many voices are heard, especially people who are less inclined to speak.

Good conversations are also the foundation of constructive civic discourse and hence a foundation of participatory democracy. A truly deliberative, consensual policy decision is a combination of iteration and improvisation. You cycle through many versions of a policy solution while searching (hopefully) for creative breakthroughs too. When people with different points of view challenge each other, they are more likely to have a breakthrough conversation if they cultivate open-minded, creative, and flexible approaches.

Tyrants, autocrats, and unilateral executives cannot tolerate dissent, and are impatient with controversy. That is why they squelch the collective imagination, promote unilateral truths, persecute those with different perspectives, and advance their own self-interests as the common good.

Truly gifted facilitators, whether in education, organizational life, or the policy arena, know how to encourage constructive conversations. They build trust among stakeholders while galvanizing creative energy. This requires a unique form of improvisational excellence: the ability to think on your feet, react in the moment, and see creative openings and let them thrive.

In *The Evolution of Imagination*, Asma discusses the importance of improvisation in the life of a moral sage, a person with supreme emotional and social intelligence. Citing the Chinese philosopher Mencius, he suggests that "the good moral improviser is not simply applying a rule to a new scenario (like a math problem), but rather she has trained her character such that she will respond to the unique complexities quickly and compassionately as an extension of her nature." Asma concludes that "a highly responsive moral virtuoso is a master improviser" who has "embodied tendencies to action, subtle perception, strong memory of successful precedents and failures, generosity beyond mere reciprocity."[37]

I believe that moral improvisation is an educational imperative for environmental learning in order to create dynamic spaces for interpreting and responding to environmental change as well as promote constructive civil discourse. That is the essence of participatory democracy. The moral improviser maintains conceptual flexibility in the face of uncertainty, is capable of facilitating conversations based on the emerging perspectives of the participants, and understands the necessity of finding a common moral ground in controversial settings. This is the challenge of integrating complex problem solving with deliberative conversations.

The *I Ching* as Improvisational Knowledge System

Remarkably, in the *I Ching* we all have access to a stream of wisdom that embodies moral improvisation within the framework of biosphere archetypes. Stretching to the dawn of Chinese history, its origins, derived from legend and folklore, are obscured by the mysteries of time, but the *I Ching* (*Book of Changes*) revolves around a comprehensive symbolism of geometric lines. There are solid (yang) and broken lines (yin), arranged into sequences of three, with each sequence representing an aspect of the biosphere. Here is the eight-trigram sequence, attributed to King Wen (1152–1056 BCE). There are other such sequences too (the book has a long history!).

These eight trigrams further combine into sixty-four hexagrams. The hexagrams consist of yin and yang lines that transform so any hexagram can morph into another one, and the interpretation of the transformation depends on whether any given line stays the same or changes. There are 4,096 permutations, all subject to infinite possibilities of interpretation.

The *I Ching* is now a remarkable wisdom collection, published in countless editions, enriched with brilliant commentaries, derived from Taoism, Buddhism, and Confucianism, and then reproduced through

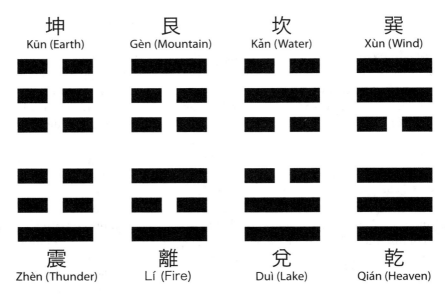

Figure 8.1
King Wen sequence

additional Western interpretations. It is unquestionably one of the world's great intellectual, ethical, and philosophical works.

You can access the *I Ching* through various divination techniques, designed to free your unconscious and enhance your receptivity to whatever situation you encounter. This is a fascinating way to synthesize an apparently random process with whatever is just under the surface of your awareness. Indeed, Carl Jung, in his famous introduction to the classic Richard Wilhelm and Cary F. Baynes translation, describes the relationship between what's random and the unconscious as synchronicity. Or you can study the *I Ching* by flipping through its pages until you arrive at a hexagram that you find particularly relevant. You can also approach it in a linear, analytic manner. Any of these methods is enriching.

There are dozens of English translations, ranging from short and poetic versions, to comprehensive, encyclopedic editions, covering multiple commentaries and interpretations. I have a bookshelf containing about twenty of these, and am always impressed with how totally different they are. Compare hexagram fourteen (Fire over Heaven) in multiple translations. For David Hinton (who offers a poetic, pithy approach), it's Vast Presence. For Jack Balkin (who emphasizes leadership and social behavior), it's Great Possession. In the Wilhelm and Baynes edition, it is Possession in Great Measure. For Deng Ming-Dao, hexagram fourteen is Great Holdings. The Rudolf Ritsema and Shantena Augusto Sabbadini translation calls it the Great Possessing.

The Ritsema and Sabbadini edition (given all the vagaries of language and interpretation) stresses long lists of evocative phrases of possibility, and you can spend several hours meditating on the nine pages of the Great Possessing. This version also contains a 170-page index, providing a comprehensive listing of the most salient words in the translation and where you will find them in the sixty-four hexagrams as well as the changing lines that connect them.

Finally, I have two books that mainly dispense with language, and instead urge readers to study the visual, symbolic, mathematical, and metaphoric relationships between the hexagrams. Thomas Cleary's *I Ching Mandalas* offers an evocative series of diagrams, excerpted from multiple Chinese philosophical traditions, designed as a study guide to better understand the intricate relationships between the trigrams. And Lama Anagrakia Govinda's *The Inner Structure of the I Ching* is a comprehensive visual guide that illustrates the geometric architecture of the trigrams and hexagrams, resulting in an astonishing array of network diagrams.

I find several translations to be exceptionally helpful, mainly because they offer compelling and evocative narratives that convey the book's ancient lineage while maintaining its relevance for a contemporary reader. *The Living I Ching* by Ming-Dao goes into great detail interpreting the eight trigrams along with discussing their origins in the natural world and the wisdom they yield accordingly. His commentary on the Earth trigram illustrates this: "It is not just the works of humans but all of nature that depends on earth. The valley is the earth flattening and opening itself to the sky. The streams and rivers run in the earth's channels. The lakes accumulate in the hollows of the earth. Without the earth, water cycles driven by heaven would not be possible. Earth gives place to the oceans."[38]

Ming-Dao takes an interesting approach to the trigrams, first exploring their meaning as biospheric phenomena, and then deriving principles of moral and ethical conduct. Of all the translations, his is the most relevant to the main focus of this section: moral improvisation and environmental learning. For example, in his discussion of the Earth trigram, he writes, "The virtues earth inspires are numerous: acceptance, support, receptivity, modesty, and perseverance."[39]

Balkin, a highly accomplished constitutional law scholar, provides a translation that underscores moral leadership and is especially useful as a guide to organizational life. Each hexagram suggests models of behavior that are contingent on ever-changing social, political, and historical situations. I frequently use his translation when I want to better understand my role in a leadership situation.

I also find John Minford's recent translation interesting as it's a synthesis of many different approaches, and he cites classic commentaries derived from multiple philosophical traditions and interpretations. Minford makes several points in the introduction that are of particular relevance to improvisational environmental learning. He describes the *I Ching* as a demanding game, explaining that you don't read it, but rather you play it, you play with it, and by doing so you maximize the quality of your interactions. Minford cites the seriousness of games, and how they represent the rituals, patterns, and symbols of human experience, embodied in how you play with the *I Ching*.[40] It's the random process of interaction that simulates a play experience.

Take a few moments and perform an internet image search with *I Ching* diagrams in the search box. You'll find charts of hexagrams as guides to spiritual drumming, the psychedelic *I Ching*, and hundreds of others. Evidently the *I Ching* has many possibilities as an improvisational knowledge system!

There's a passage in Minford's translation that best exemplifies the *I Ching*'s origins as a nature-based oracle. In the introduction, he recounts that in doing research for the translation, and studying the copious and intricate commentaries as well as the historical research, he found himself asking what he portrays as the translator's first question: Which *I Ching* am I translating? Is it the cryptic ancient *Oracle* or the more elaborate, historically contextual *Book of Wisdom*? He concludes that it is both, and his translation will reflect that, but he also realizes the necessity of respecting its ancient origins. I quote him at length here because it demonstrates how the moral depth of the *I Ching* is ultimately a reflection of studying the biosphere:

> Where was Fu Xi, the caveman-sage wrapped in his furs, contemplating the origins and mysteries of the Universe, inventing the Eight Trigrams to make some sense of it all? In the words of the Great Treatise . . . he gazed upward and observed Images in the Heavens (the "night sky"—he was an astronomer); he gazed about him and observed Patterns upon the Earth (he was a geologist, a geographer, a geomancer). He observed markings on birds and beasts (he was a naturalist), how they were adapted to different regions (he

was an ecologist).... He drew inspiration from within his own person (he was a psychic, a psychologist); further afield, he drew inspiration from the outside world (he was an empirical scientist).[41]

This passage beautifully connects the potential correspondence between environmental observations and human wisdom, and reveals the extraordinary potential within the *I Ching* as an improvisational biosphere knowledge system. In the twenty-first century, as environmental change threatens ecosystem integrity, we desperately require ethical and moral wisdom. The *I Ching* supplies an interpretative model, potentially linking ancient wisdom with environmental change science. In the next section, I'll suggest some educational approaches for exploring these possibilities.

An Environmental Change *I Ching*

Let's return to hexagram fourteen (Fire over Heaven) and use Minford's descriptor, Great Measure. His interpretations of this hexagram focus on leadership in a time of prosperity, the importance of promoting what is good in an organization (or social setting, or state), and the necessity of gaining the respect and support of colleagues, while being wary of the pride and arrogance that often follows prosperity, potentially leading to misfortune.[42] Balkin's interpretation stresses the necessity of remaining modest and unassuming. "If you are magnanimous, generous, and humble, you can make great progress, gather increased support, and draw even more resources to your side."[43] This hexagram is a wonderful meditation on generosity.

What if we interpret the hexagram from an environmental change perspective? And then combine that interpretation with the classic narratives? What image do you conjure when you consider Fire over Heaven? I immediately envision the heating of the sky, a warming atmosphere, increased energy in the biosphere, and the various climatic changes that may result. I imagine the prosperity of Great Measure as the increased energy of economic activity. The hexagram warns against the hubris that frequently follows prosperity. I think about fossil fuel companies that enrich themselves at the expense of the planet, manipulate civic conversation to achieve their aims, and how that represents an abuse of prosperity.

I appreciate that this may seem improbable. And of course, I mean no disrespect to an extraordinary book that integrates several thousand years of philosophical traditions, cultural relationships, and languages. I believe that it is in the spirit of the *Book of Changes*, however, to consult the oracle and engage its wisdom in whatever manner meets the needs of our time. Can we create an environmental change *I Ching*—one that uses the biosphere-based trigrams as a foundation, and then incorporates the centuries of wisdom in guiding action and response?

There are many ways to approach this. You can start by renaming the hexagrams based on the biosphere metaphors they convey, such as Wind over Water (Species Migration), Earth over Fire (Plate Tectonics), and Water over Mountain (Hydrologic Cycle). There will be sixty-four combinations in all—surely enough for a stunning variety of environmental change processes. Think, too, about the relationships of changing lines. What does it mean when Water over Mountain transforms to Fire over Mountain? Does that imply the danger of wildfires? With every new hexagram, you can write interpretations that propose human action in response to ecological change, from stewardship advice to civic responsibility.

This would be an excellent teaching activity, working with groups of students, let's say in an environmental studies class, to consider these arrangements, and then develop their own metaphors, images, responses, and wisdom. You can develop unique biosphere images or metaphors based on prevalent patterns in nature, and link them to the hexagrams so as to be incorporated with outdoor activities. In cities, you could ask participants how they encounter the eight biosphere archetypes, and how those archetypes impact city life. You could assign a specific hexagram to several groups and compare their interpretations. Over the course of a semester, you could construct a workbook of possibilities, culminating in diagrams, artwork, engaging narratives, and storytelling, organized around the idea of an environmental change *I Ching*.

I'm suggesting that we use the best empirical data from environmental change science as the basis for imagery, metaphors, and narratives that enhance our ability to observe, interpret, and then respond to environmental change. The *I Ching* is sufficiently resilient that it may be a template for such a project, either using the traditional biosphere trigrams or developing alternative pattern-based systems.

These approaches can be coordinated with what is already an impressive environmental education movement based on immersive, place-based outdoor learning. The Forest kindergarten school network encourages children to look at patterns in nature, and use that exploration for teaching principles of ecology as well as empowering students to engage in what it describes as flow learning—an approach that integrates play, imagination, and empirical inquiry. David Sobel explains in "Outdoor

School for All: Reconnecting Children to Nature" how this emphasis on exploratory, play-oriented, immersive outdoor learning is one element of a global movement, throughout the K-12 spectrum, supporting a place-based orientation, with curricular approaches in geographies as widespread as the Australian bush, coast of British Columbia, and inner-city Saint Louis. Improvisational learning in outdoor settings is crucial to this movement.[44]

What's most inspiring about the *I Ching* as a template for environmental learning is how it further stimulates improvisational thinking. The conditions of our learning are always changing, and the best teachers understand this, providing ways for students to adapt to a variety of learning environments and multiple ways of knowing. They guide them in learning how to integrate those skills with a moral foundation based on ethical considerations. This chapter describes how this conceptual flexibility is enhanced by improvisational excellence. In the final section, I'll take this one step further. I'll suggest that improvisation ultimately is a way for us to emulate and express gratitude for the biosphere, which makes our lives possible.

Does Improvisation Emulate the Biosphere?

Go for a walk; cultivate hunches; write everything down, but keep your folders messy; embrace serendipity; make generative mistakes; take on multiple hobbies; frequent coffeehouses and other liquid networks; follow the links; let others build on your ideas; borrow, recycle, reinvent. Build a tangled bank.
—Steven Johnson, *Where Good Ideas Come From*

It is interesting to contemplate a tangled bank, clothed with plants of many kinds, with birds singing on the bushes, with various insects flitting about, and with worms crawling through the damp earth, and to reflect that these elaborately constructed forms, so different from each other, and dependent upon each other in so complex a manner, have all been produced by laws acting around us. . . . There is grandeur in this view of life, with its several powers, having been originally created into a few forms or into one; and that, whilst this planet has gone cycling on according to the fixed law of gravity, from so simple a beginning endless forms most beautiful and most wonderful, have been, and are being, evolved.
—Charles Darwin, *The Origin of Species*

These books, separated by 150 years of ecological and evolutionary intellectual development, demonstrate the power of a connective metaphor. The tangled bank, one of Darwin's most enduring metaphors, indicates a reverence for the magnificent intricacies, possibilities, and combinations of the biosphere, intimating that the evolutionary process is the ultimate creative impulse. Johnson, in his excellent book *Where Good Ideas Come From: The Natural History of Innovation*, suggests to his readers that to stimulate their creative energies, they should build a tangled bank of interests, initiatives, and networks. Both passages are the final statements (the last paragraph) of the books. Both are testimonies to improvisational excellence.

I am constantly drawn to another great book that serves to further enhance these passages. In *Free Play*, Nachmanovitch reflects on the evolutionary origins of improvisation somewhat similarly: "An improviser does not operate from a formless vacuum, but from three billion years of organic evolution; all that we were is encoded in us."[45]

For decades, environmental activists have argued against a view of nature that regards the biosphere as another kind of bank: a repository of natural resources, vast oceanic and atmospheric field for depositing waste, or global landscape to be divided into enclosed territories of economic development. Even sustainable development, for all its merits as a judicious and practical ecological initiative, embodies this view, however tempered it may be. And on a planet now approaching eight billion people, the necessity of resource extraction is ubiquitous. A tangled bank indeed!

Yet there is another compelling view—one that brings a sense of balance, reciprocity, and gratitude to human agency. This view emerges from many ancient and modern traditions, both contemporary and indigenous, intellectual and practical, aspiring and humble. It promotes human flourishing while offering praise to the biosphere processes that make our lives possible. It understands the necessity of creativity, the arts, the spiritual, and the mysterious. It acknowledges that human life is constantly finding ways to overcome the limits that constrain it. It promotes adaptive flexibility and improvisational excellence as a means of enriching the spirit as well as creating a sustainable economy and culture.

I believe that the biosphere (nature) is the original improvisational milieu. The three billion years of organic evolution encoded into us represents an incomprehensible chain of intricate possibilities. When we improvise, we emulate biosphere processes, ecosystem relationships, and the great swirl of life as it explores the vast possibilities for continuity and survival. As we cultivate a deeper awareness of those processes, we are more capable of dwelling in their flow, enriching our experience, and understanding the human place in the here and now. As sentient manifestations of the biosphere, we are the living embodiment of improvisational process. It comes naturally to us. And when we improvise, notwithstanding its adaptive function, we are emulating the world from which we came. The biosphere responds. Call and response. We sense the biosphere's grandeur. We penetrate its mystery. We pursue the unending possibilities of the interface between mind and ecosystem. We observe the life histories of the magnificent biodiverse species with which we cohabit the planet. And from those observations along with our deepening understanding of environmental change, we develop creative solutions for human flourishing in the biosphere.

9

Perceptual Reciprocity

A Glimpse through the Fog

During summer 2018, the Monadnock region of southwestern New Hampshire was unusually warm. We had a series of hot and dry days in late June and early July, followed by a long stretch of humid weather, as a southwesterly flow took hold of the region. One of my favorite summer activities is to take long swims in Dublin Lake, a spring-fed, 236-acre body of water nestled against Mount Monadnock, a 3,166-foot peak. It's a beautiful lake, with restricted motorboat access, and if you take your swim in the early morning, you're almost always alone. That summer, the lake was sufficiently warm that I could comfortably take sixty- to ninety-minute swims.

I'm one of the slowest swimmers on the planet. I alternate between the breaststroke and sidestroke, while keeping my head above water. So these swims are really long hikes, except I'm in the water. My eyes are just above the water level. I gaze across the surface of the lake while still being able to raise my eyes to take in the whole lake and/or sky. Over the course of the summer, I take great pleasure in the magnificent variety of lake and sky conditions, seasonal changes, temperature and movement of the water, patterns of insects, and lovely shades of blue, green, and gray, as the water and sky become a recursive, infinite mirror of light and sound.

One August morning, just after dawn, on a particularly humid day, I took a memorable swim in the early morning fog. The water was nearly still, in contrast to the gentle flow of fog on its surface. I could barely make out the shore just one hundred feet away. Over the next hour, the fog would ebb and flow. As it slowly began to lift, I noticed multiple layers of fog and clouds—the white, wispy clouds on the surface, the dense layer above that covering half the sky, the sun appearing like a lost moon, barely shining through that layer, and then the holes in that layer revealing yet a third layer of altocumulus clouds, and through them a few peeks of blue sky. In contrast to the relatively still water, the atmosphere was vibrant and dynamic, with the sun slowly burning and dissipating the multiple layers of fog and clouds.

On closer observation, the seemingly still surface of the water revealed subtle shifts. Just as I approached the end of a narrow peninsula, as I left a sheltered cove, I noticed a demarcation between the utterly calm, smooth surface and gentle ripples around the bend. On the surface of the water, I made my way through a maze of water striders, forming what seemed like multiple triangular patterns as they skipped along the surface. And about five feet above the water striders, dragonflies and damselflies scurried over my head. For about five minutes, about fifty feet away, two loons floated along, and finally dived under the surface. They never reappeared.

The coastline features trees that hang over the water, white pines, hemlocks, and some oaks and maples as well as blueberry and rose bushes. At first the fog covered the tips of the trees, shrouding them in mystery. As the fog lifted and more sunlight came through, the colors became brighter. I noticed the first glimpses of yellow breaking through the green—a sure sign of late summer.

A glimpse. The etymology of "glimpse" is "shine faintly, transient appearance, a brief and imperfect view."[1] What an evocative word. It conveys the idea that there is something beyond what you see and know—a deeper layer, a potential, a world beyond your ordinary experience. For a moment something is revealed, but the mystery is never fully accessible. A glimpse can be a sign or even warning. But mainly it's a beckoning, a goad to curiosity, a portal to deeper awareness, a signature of the ineffable, and an opening for learning.[2]

During that memorable morning swim, there were moments when I was fully engaged. Rather than seeing myself as a separate body moving through the scene, I experienced a seamless continuity. But those moments were fleeting. There were brief stretches when I enjoyed a totality of awareness, a genuine feeling of interdependence, when I viscerally understood that I was one perceiving player among many. More typically my experience was limited by my identity as a human observer, embodied actor, and discrete self, desperately trying to take it all in, and in my desperation to transcend that perspective, I was further separating myself from the extraordinary possibilities of the experience.

Here's the core of my challenge. Perhaps you experience something similar. Throughout my years of reading, thinking, writing, and educating about expanding ecological awareness, I mainly grasp ideas such as species interdependence, environmental change, and coevolution as intellectual constructions. I get it analytically. I know why these concepts are important and should be fundamental to environmental learning. Yet only rarely do I enter the expanded awareness of a true flow experience, a deeper emotional or even spiritual encounter with the landscape along with the expanded community of species and the biosphere. The glimpses are inspirational and may sharpen the intellect, but a glimpse is all I get, or perhaps all I am capable of or even allowed.

For my entire life, I've tried to use intellectual experience to prepare me for a deeper awareness of organisms, landscapes, and habitats. In large measure, that is a satisfying and stimulating path. It's not enough, however, and I know there is something more. How do I balance the grasping quality of intellectual stimulation with a bolder penetration into those glimpses of wonder?

For decades, this has been the overwhelming challenge of environmental learning: to expand awareness of the natural world, form a deeper communion, and in so doing, build respect, gratitude, and appreciation for the conditions of our human existence on a magnificent planet. Here's the rub. Amid all the distractions, traps, and frailties of the human experience, that level of awareness and gratitude is exceptionally difficult to achieve. But without it, environmental learning lacks soul and passion, and it will never be more than an intellectual accomplishment. What kind of educational process allows us to achieve a balance between intellectual growth and a penetrating gratitude for the biosphere where we dwell? What kind of visceral experiences promote that balance?

Throughout this book, I've emphasized the necessity of enhancing perceptual experience. And that perception, observation, and interpretation are the necessary ingredients for expanding environmental learning. There's a fourth word that extends this sequence: reciprocity. It conveys sharing, mutuality, movement back and forth, an oscillation or wave of interconnected, reflexive experience.

I propose that we combine perception and reciprocity to form perceptual reciprocity. What does that expression convey? I suggest that all living things on the earth perceive. I'll go beyond that and propose that there are collectivities of perception. Communities of species perceive. Forests perceive. Habitats perceive. Landscapes perceive. The biosphere perceives. We should think of perception not simply as a specific moment of awareness but rather a collectivity of awareness in multiple measures of place and time. Perceptual reciprocity asserts that these collectivities of awareness inform each other. To apprehend environmental change requires a deepening of perceptual reciprocity.

In this chapter, I'll explore what that means and how we might come closer to achieving it. I'll describe a compelling approach for expanding perception—a theory of the entropic mind—and its educational implications. This includes a discussion of the *umwelt* concept: a method for imagining the sensory experience of other creatures. How do animal encounters expand our umwelt? I'll introduce additional perceptual modalities—how a forest perceives—and then add another layer—the awareness that our bodies are conglomerations of multiple species. What we typically conceive of as self and identity is much more ecologically as well as psychologically intricate. This leads to a discussion of memory and how the earth reveals its history. Tapping into that history is another way to enhance perception. I'll suggest that perceptual reciprocity might begin in your neighborhood, the local place that you know best. I'll conclude by adding one more word to our sequence: generosity. It catalyzes observation, perception, interpretation, and reciprocity.

The Entropic Mind

When I returned from that inspiring summer swim and the glimpse through the fog, I retreated to my study. I had been reading Michael Pollan's remarkable book *How to Change Your Mind*. As Pollan enters his sixties, he ponders how frequently he ignores the present moment, always "jumping ahead to the next thing." Pollen observes, "Alas, most of the time I inhabit a near-future tense, my psychic thermostat is set to a low simmer of anticipation and too often, worry. The good thing is I'm seldom surprised. The bad thing is I'm seldom surprised."[3]

Furthermore, he worries that his routine habits, although they have been incredibly effective at creating an accomplished, nourishing, and well-balanced life, also diminish his creative potential. "Habits are undeniably useful tools, relieving us of the need to run a complex mental operation every time we're confronted with a new task or situation. Yet they also relieve us of the need to stay awake to the world: to attend, feel, think, and then act in a deliberate manner."[4]

He's striving to find ways to expand his present-moment awareness, both to better understand the nature of consciousness and deal with some of the existential challenges of aging. Pollan is searching for approaches to help him break out of these routines. He is intrigued by the prospect of a psychedelic path, although he is aware that "psychedelics are certainly not the only door to these other forms of consciousness . . . but they do seem to be one of the easier knobs to take hold of and turn."[5]

How to Change Your Mind is a comprehensive, highly personal, and absolutely riveting journey through the history and politics of psychedelics, including visits with some of its enthusiastic advocates and a review of the neuropsychological research surrounding their use, including comments on their therapeutic potential. Pollan reports on his own tripping experiences, the mentoring he received, and the extent to which they in fact helped change his mind.

This book checked many boxes for me. We are roughly the same age, come from similar cultural backgrounds, and were both highly influenced by the 1960s. I also feel limited by my habitual routines and inhabitation of the near-future tense. I greatly respect Pollan's environmental writing, especially *The Omnivore's Dilemma*, a brilliant survey of the US food system, and *The Botany of Desire*, a wonderful exploration of the coevolutionary relationships between humans and plants. There's a terrific chapter in *How to Change Your Mind*, "Natural History," that brings the same coevolutionary perspective to his discussion of psychedelics.

In that chapter, Pollan spends time with Paul Stamets, a mycology expert who believes that psychedelic mushrooms are a doorway to an enhanced collective consciousness of the biosphere. Pollan believes that Stamets is extending the tradition of the Romantic scientists, among whom he includes Humboldt, Johann Wolfgang von Goethe, Joseph Banks, Erasmus Darwin, and Thoreau.[6] "Instead of seeing nature as a collection of discrete objects, the Romantic scientists—and I include Stamets in their number—saw a densely tangled web of subjects, each acting on the other in the great dance that would come to be called coevolution," Pollan writes. "'Everything,' Humboldt said, 'is interaction and reciprocal.' They could see this dance of subjectivities because they cultivated the plant's eye view, the animal's eye view, the microbe's eye view, and the fungus's eye view—perspectives that depend as much on imagination as observation."[7]

What does it take to achieve these different perspectives? I'll examine this question throughout the chapter and hopefully for the remainder of my life too. Before I do, I want to cover what I think is another relevant aspect of Pollan's book: the research that explains how as well as why we open and close our minds.

Pollan uses a diagrammatic metaphor to capture brain activity under psilocybin (see figure 9.1). Then he links this conceptualization of brain network expansion to evolutionary adaptation:

One way to think about this blooming of mental states is that it temporarily boosts the sheer amount of diversity in our mental life. If problem solving is anything like evolutionary adaptation, the more possibilities the mind has at its disposal, the more creative its solutions will be. In this sense, entropy in the brain is a bit like variation in evolution: it supplies the diversity of raw materials on which selection can then operate to solve problems and bring novelty into the world.[8]

The entropic brain hypothesis suggests there are a great variety of mental states in which the networked, synaptic connections in the brain proliferate. Pollan cites the groundbreaking work of Robin L. Carhart-Harris and his research team, reviewing their technical paper "The Entropic Brain: A Theory of Conscious States Informed by Neuroimaging Research with Psychedelic Drugs."

This theory hypothesizes a spectrum of cognitive states. It asserts that in high-entropy mental states, the mind is highly flexible and cognitively engaged, but also highly disordered. Examples of such states are dreaming, magical thinking, creative improvisation, and psychedelic experiences. In low-entropy states, the mind is more rigid yet much more ordered. Examples include addiction, obsessive-compulsive disorder, and depression. Of course, these are ideal types. They vary according to cultural environments and the idiosyncrasies of personality. Nevertheless, the scheme is intriguing and especially helpful for how we might think about environmental learning.

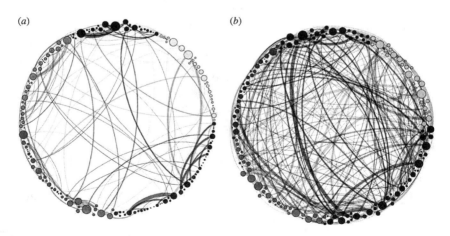

Figure 9.1
Brain activity under psilocybin. Courtesy of Petri et al., "Homological Scaffolds of Brain Functional Networks."

When I am most engaged, my mind is highly flexible and creatively involved; my eyes and mind are wide open. When I experience obsessive-compulsive disorder tendencies, I falsely assume that by controlling several variables, I can create order in my life, fend off uncertainty, and limit risk. Most waking consciousness exists in between those extremes. You dramatically limit your creative potential when you mainly dwell in low-entropy, high-order states, or what I'll call the closed-mind region. And if you only exist in the region of high disorder, you risk losing your ability to function in ordinary reality.

Pollan discusses the relativity of experience and how our "normal waking consciousness" may be "less a window on reality than the product of our imaginations—a kind of controlled hallucination." He notes how all mind states on the Carhart-Harris and colleagues' spectrum are informed by our sensory observations, and then sifted through the imagination. "But in the case of normal waking consciousness, the handshake between the data of our senses and our preconceptions is especially firm." And indeed it must be because "ordinary waking consciousness has been optimized by natural selection to best facilitate our everyday survival."[9]

As an environmental learner, I'd like to find ways to spend more time in the open-mind region and do so in ways that don't stray too far from waking consciousness. Is there perhaps a middle ground between waking consciousness and high entropy? And are there ways that educators can facilitate guided approaches to that middle ground?

I want to be clear that I am not advocating the use of psychedelics to enhance perceptual reciprocity, although I do think turning that knob holds fascinating potential. What I am suggesting is that the process of "changing your mind" or enhancing your perception is crucial for environmental learning, and the neuropsychological research around psychedelics may yield helpful insights. There are many other ways to get there, such as forms of meditation, participation in flow-filled improvisations, or immersive outdoor experiences. The challenge is learning to dwell more fully in the present moment, and by so doing, enhancing your ability to observe the world to establish perceptual reciprocity with other species, landscapes, and the biosphere.

We could try what Pollan depicts as "another trippy thought experiment" and imagine the world from the perspective of another creature who relies on an entirely different sensory apparatus—let's say, the octopus, with its radically decentralized brain and intelligence distributed across eight arms. Let us add the octopus's "arm view" to the dance of subjectivities cited earlier.

This is what many children's book do. And it's also what many scientists do when they study animal behavior. They become so immersed in the lives of the creatures they study that they develop the empathy and imagination to sense the world from the perspective of the creature. In 1934, Jakob von Uexküll proposed the term umwelt to portray how we might engage with another creature's perspective. His classic work *A Foray into the Worlds of Animals and Humans* describes some methods for doing this and thus is an essential approach for perceptual reciprocity.

Floating Bubbles

But what if I told you that it is very likely that whales communicate with other whales over miles of open ocean? Or that many animals have an internal compass that they use as a navigational tool that permits them to migrate over long distances? Or that some fish communicate using coded messages that are sent though the water via electrical fields, and that bees and other insects see things that are completely invisible to our eyes?
—Howard C. Hughes, *Sensory Exotica*

I'll never forget the first time as a child that I discovered a diagram of the electromagnetic spectrum. I saw the vast array of waveforms, and how they were mainly inaudible and invisible. What I could hear and see was just the thinnest cross-section of possibility. Around the same time, my younger brother became fascinated with fluorescent rocks and minerals. He acquired an ultraviolet light that illuminated otherwise-bland rocks into a mind-blowing display of color. I was equally mystified by the fact that invisible radio and television waves as well as gamma rays from the sun along with zillions of other invisible waves were passing through my body. Other animals could sense some of those wave pathways, but they couldn't necessarily see the same colors I could. There's a lot going on out there, and I could only viscerally experience a small part of it. What a narrow view! And yet I was enamored with the extraordinary depth of visual, auditory, and olfactory experiences that were within my range, and fully aware I was only taking in a small part of what my senses had to offer.

Every creature that lives on this planet has a unique sensory array, developed through evolutionary adaptation and linked to a survival strategy. In fact, the visible, olfactory, and auditory spectra that define human sensory awareness are unique to our species, and different individuals experience a range of perceptual possibilities.

Von Uexküll introduces the umwelt concept by asking his readers to imagine a stroll on a sunny day in a flowering meadow surrounded by insects and butterflies. He suggests that the reader create a bubble around each of the creatures dwelling in that meadow. That bubble is a metaphor that circumscribes any creature's perceptual world. "As soon as we enter into one such bubble, the previous surroundings of the subject are completely reconfigured. Many qualities of the colorful meadow vanish completely, others lie there incoherent with one another, and new connections are created. A new world rises in each bubble."[10]

If you want to learn how to change your mind, imagine entering one of these bubbles—let's say the bubble that surrounds a red-tailed hawk. How would you do so, and what would you learn? It's challenging to do this. The amateur naturalist engages in hours of observations. The research scientist organizes detailed experiments. The artist conjures the imagination. The shaman may utilize enhancements such as hallucinogenic mushrooms or plants, sensory deprivation, or other forms of dreamlike trances. All these approaches—observation, experimentation, imagination, and altered states—may enlighten how you perceive the world through that bubble, but you will forever be limited by your own organismic sensory capacities. Entering the bubble is a way to enhance those capacities given those limitations. How wonderful it would be to experience the world (even for just a few moments) as a red-tailed hawk or whatever creature fascinates you.

There are many epistemological and philosophical questions inherent in this methodology. Von Uexküll's work is considered an important foundation of biosemiotics, the study of signs and codes in the biological realm, with implications for ethology (the study of animal behavior), biological communication, and cognitive philosophy. In a lengthy introduction to von Uexküll's work, Dorion Sagan explains that the idea of a "distinct perceptual universe" expands our understanding of nature beyond the realm of the human. Sagan writes that Uexküll's vision reminds him of Indra's net, a Buddhist concept describing "an infinite web with a dewdrop-like eye glimmering in the middle of each compartment. Each jeweled eye contains all the others and their reflections."[11]

Indra's net is a beautiful concept. It proposes an infinite interconnectedness of consciousness. Each perceiving being is connected to the whole network, although it only apprehends that network from its particular perspective. Thich Nhat Hanh refers to this as "interbeing": the idea that "everything relies on everything else in the cosmos to manifest—whether a star, a cloud, a flower, or you and me."[12]

These self-enclosed, discrete perceptual bubbles are interconnected, although these connections may not be evident. The red-tailed hawk that floats on the thermals high above me can certainly see that I am sitting here, but I am not something it can eat, and I barely register in its perceptual field. Nor can the red-tailed hawk possibly know that I am writing about it to explain umwelt, Indra's net, and perceptual reciprocity—terms that may help me understand my relationship to the hawk, yet are clearly enclosed within my conceptual bubble and very much a human way of thinking.

Picture yourself again in von Uexküll's meadow and imagine all the creatures as bubbles, floating through space, sometimes interacting or even colliding, and at other times drifting along without any encounters. Each of these floating bubbles is a discrete yet connected perceptual universe, and the relationships of these bubbles, how and whether they encounter each other, is characteristic of the meadow ecosystem. Every landscape has a different configuration of species (hence bubbles) and an almost inconceivably complex assortment of perceptual possibilities.

The challenge for environmental learning is how to change our minds so we can better appreciate the magnitude, intricacy, and connectedness of these floating bubbles. Why is this important? This kind of expansive thinking places our limited human experience in a broader ecological as well as biospheric context. If this seems too improbable or vaguely philosophical, let's bring it back to the most basic terms. Check out Sy Montgomery's wonderful book *How to Be a Good Creature*. She essentially writes a memoir describing how her life has been immeasurably enhanced by her "friendships" with thirteen different species, including a tarantula, tiger, pig, and octopus. It doesn't have to be complicated. You just learn how to come out of yourself enough to share a common, if different, perceptual space—the foundation of interspecies communication. You don't need psychedelics to open the doors of perception. Sometimes an intimate relationship with another creature will do just fine. Interspecies communication is a first step toward perceptual reciprocity.

Becoming a Good Creature

What have animals taught me about life? How to be a good creature.
All the animals I've known—from the first bug I must have spied as an infant, to the moon bears I met in Southeast Asia, to the potted hyenas I got to know in Kenya—have been good creatures. Each individual is a marvel and perfect in his or her own way. Just being with any animal is edifying, for each has a knowing that surpasses human understanding. A spider can taste the world with her feet.

Birds can see colors we cannot begin to describe. A cricket can sing with his legs and listen with his knees. A dog can hear sounds above the level of human hearing, and can tell if you're upset even before you're aware of it yourself.
—Sy Montgomery, *How to Be a Good Creature*

Montgomery's best teachers are animal companions. Her many books are vivid journeys through which she engages in the lifeworlds of an extraordinary menagerie of animals, both close to home and in exotic places. *How to Be a Good Creature* is a personal, riveting, and humble memoir of these journeys—a developmental autobiography told through animal encounters.

Her life narrative is both an environmental learning odyssey and pursuit of virtue. The animal encounters require careful observation, courage, tenacity, boldness, and empathy. When she enters their habitats, her challenge is to prove she poses no threat, and is trustworthy and respectful. In many cases, most notably with "Octavia," the octopus, she develops friendships, or at least signals of recognition and joy, playful meetings and connections, that forge social as well as emotional bonds. As you read about Montgomery's experiences, you realize how much these creatures teach her about life and herself; the emus, for example, taught her patience, independence, and perseverance. But what about the rest of us? What if you're uncomfortable around animals or can't journey off to Australia to study emus? Or you'd rather not get to know the local tarantula?

Many people at some point in their life develop an intimate relationship with a dog or cat. I don't wish to generalize as to the nature of those relationships because they are as variable as the people and animals involved. It's safe to say, however, that people often develop intimate, caring relationships with their pets, and if they are open to better understanding how their pets observe the world, there's much to learn about their pet's and own behavior. If you take your dog for a walk and don't try to constantly control the leash but instead immerse yourself in the dog's sensory path, your walk will go to different places than if you were on your own.

We once had a small dog, a schnauzer named Poncho, who was free to roam the woods around our home in rural New Hampshire. Poncho would disappear for hours at a time, but would always make his way home, sometimes at two in the morning. At home, he was a loving, cuddly creature, but in the woods, he became a wild animal, stalking deer, running with other dogs, and eventually meeting his demise in an unfortunate and sad scuffle with a coyote. Our next dog, Paco, also a schnauzer, was a good hiker, but he would never stray too far from his human pack. Paco loved to play ball. I once observed him with a stick in his mouth, entirely on his own with no human "supervision," running around our driveway path and hitting a ball with that stick. Dogs, of course, remind us, if we are so willing, of our own animal bodies. More frequently than not, dogs are good creatures and bring out the best in our character.

When Paco was old, I was impatient with him on walks, and on several occasions yanked hard on his collar when he didn't want to proceed. Subsequently I felt ashamed of doing so, observing that his old bones undoubtedly needed a rest. I realized that Paco's old age was a precursor to my own and he had a great deal to teach me about aging. Now that I am the same age (in so-called dog years) as he was when I did that yanking, I understand how impatient and insensitive I was.

This may seem like a trivial example. But I think most of us can recall profound learning experiences we've had in our relationship with an animal. In the spirit of full disclosure, I will tell you that if I encounter an animal in the wild, I have no idea how I will respond. When I see a rat on a city street I am repulsed. I always avoid handling snakes when I visit nature centers. Yet when I encountered a poisonous tiger snake on a trail in Tasmania, Australia, I had no fear, only the requisite caution and curiosity. When I was visiting Cambodia and a young girl put a tarantula on my chest (in hopes of selling it to me), I was uncomfortable though surprisingly calm.

I bring up these instances because I think the greater challenge is how to develop respect and humility for all creatures, even those who might prefer to eat you. You might not want to hang out with elephants (or maybe you do), but you can still be in awe of their emotional and social intelligence. You can still read about them to expand your understanding of the world around you.

Consider, too, the path that Rothenberg takes in relation to interspecies communication. Rothenberg searches for musical common ground. He places himself in environments that allow him to play his clarinet with birds or establish musical connections with whales. His musical strolls or forays with these creatures are superb illustrations of perceptual reciprocity. There's much to learn from birdsong and whale song alike.

There are countless paths to interspecies communication, including those that result in the death of the animal, such as hunting relationships. In *The Island Within*, Richard Nelson writes about how crows and ravens have been known to pass along tips to indigenous hunters in pursuit of deer.

I have no doubt that these circumstances—communicating with octopuses, playing with your dog, jamming with whales, or hunting deer—are conducive to opening the human mind. And that if you are in the right frame of open mind, there's much to learn about becoming a good creature.

There's an important caution here as well. We are all susceptible to anthropomorphizing: projecting our own human emotional perspectives onto the animals we think we communicate with. I love Bambi too, but I'm not sure I gain much from thinking about animal encounters this way, as heartwarming as it may be to my grandchildren.

Animal encounters are ubiquitous and yet still mysterious. It's so easy to forget we are animals too. So many species of animals—especially yet not exclusively the charismatic megafauna—are threatened. The only conceivable way to reverse that trend is to develop the empathy and compassion to better understand why these animals matter. Reading vivid accounts of animal lives is a helpful start. In *Beyond Words: What Animals Think and Feel*, Carl Safina provides detailed accounts of the emotional lives of elephants, wolves, and whales. Most important, he emphasizes that a deeper understanding of these animals challenges how we think about intelligence, cognition, and emotion. Safina urges us to open our minds to all the intelligent life on the planet.

In *Zoologies*, Alison H. Deming writes a series of short essays on animals both common and obscure—from oysters to vultures to rabbits—in her effort to better understand what it means to be human, and in so doing, give more meaning and perspective to her life. "I have a strong attraction to animals, and I sometimes think that my animal self knows that I am of their world, that they have claimed me, humbled me, and left their mark on my body and spirit. I feel I owe them my attention."[13]

Perceptual reciprocity means we do our best to learn from our animal encounters, whether it's mice or elephants, mosquitoes or dolphins, instead of informing them with a human perspective. That's one crucial way to become a good creature.

The Great Connectors

To claim that forests "think" is not an anthropomorphism. A forest's thoughts emerge from a living network of relationships, not from a humanlike brain. These relationships are made from cells inside fir needles, bacteria clustered at root tips, insect antennae sniffing the air for plant chemicals, animals remembering their food caches, and fungi sensing their chemical milieu. The diverse nature of these

relationships means that the tempo, texture, and mode of the forest's thoughts are quite different from our own.
—David George Haskell, *The Songs of Trees*

Who does the tree-hugger really hug, when he hugs a tree?
—Richard Powers, *The Overstory*

As mysterious as they are, animal encounters are visceral, tenable, and familiar. You can hug your dog (though maybe not a rat) and receive an emotional response. Surely, too, you can hug a tree, but you're unlikely to get a tangible emotional response. Nevertheless, people develop powerful relationships with trees. In his great environmental novel *The Overstory*, Powers writes about the protesters who inhabited tree houses in the redwoods and their dramatic efforts to halt logging. He traces the life stories of the protesters and how they developed such powerful emotional relationships with trees—enough to place their lives in jeopardy to save them. Powers intertwines the life stories of a dozen families in the United States with their relationships to trees, retelling US history as the interconnected lives of people and forests. The protesters come from many directions and backgrounds. There are working people, scientists, thrill-seeking college students, engineers, and people who are lonely or otherwise abandoned. Their life circumstances allow them to bond with specific trees or form friendships with other tree lovers, and through those bonds, develop emotional connections with all trees everywhere.

Powers develops a parallel narrative in *The Overstory*, mainly through his portrait of an important protagonist: a botanist whose research brings to life the role trees play as nature's great connectors. In a courtroom setting, Patricia, a scientist, tells the judge why the old-growth redwoods shouldn't be logged. Powers observes that "love for trees pours out of her—the grace of them, their supple experimentation, the constant variety and surprise. These slow, deliberate creatures with their elaborate vocabularies, each distinctive, sharing each other, breeding birds, sinking carbon, purifying water, filtering potions from the ground, stabilizing the microclimate. Join enough living things together, through the air and underground, and you wind up with something that has *intention*. Forest. A threatened creature."[14]

Haskell covers similar ground in *The Songs of Trees*, subtitled *Stories from Nature's Great Connectors*. He writes about thirteen different tree species—ceibo, balsam fir, sabal palm, green ash, mitsumata, hazel, redwood, pondersosa pine, maple, cottonwood, callery pear, olive, and japanese white pine—exploring their cultural and natural histories, ecological functions, aesthetic appeal, and spiritual significance. He also offers a sobering reminder of the threats to global forests. He examines every nook and cranny of trees, from the topography of the canopy to the intricate ant/fungal/bacterial convergence of lives that they host as well as support. In describing the crown of the ceibo, a tree in the Ecuadoran rain forest, he observes "there are as many microclimates among those branches as exist within hundreds of hectares of most tropical forests. Bogs accumulate in sheltered crotches. Ephemeral wetlands fill and dry within knotholes.[15]

Haskell covers the many functions of trees as great connectors—photosynthesis, biogeochemical cycling, and hydrologic stabilization along with the vast underground connections of mycelia, soil microorganisms, and root systems. Most of these processes occur "at a scale that evades our senses." These processes are embodied networks—vast series of connections that take place within and between trees as well as the surrounding environment, between trees and an entire matrix of organisms, and between trees and ourselves. To recognize and appreciate these connections requires both scientific study (Haskell has done much of that) and full immersion in the lives of these trees, such as being in their presence, observing them over time, studying their relationships, and considering what they can teach us about being a good creature.

To return to Powers's question, What does the tree hugger really hug? What is the tree hugger embracing? I just got back from a walk on a cold, late December winter day in the oak and maple forest of southwestern New Hampshire. After a spell of rain, the landscape is bare, and the November snows have all but disappeared from the forest, surely to return in the months to come. I take a break from my writing to literally walk the talk. I decide to hug one of the largest and oldest trees, a tall eastern white pine. On the surface, it seems lifeless. The insects are moribund. I glance toward the canopy and notice the dark green needles against the backdrop of the blue sky. A flash of color on a brown day. The soil activity is mainly frozen. Most of the birds have gone south. There is some squirrel activity, but mainly they've retreated to cavities and nests. The trees appear to be at rest, taking a break (or so it seems) before the rapid biomass production that will take place between May and July. There are thousands of trees in this forest. If I hadn't read it in a book, I wouldn't necessarily know they are in constant communication. They dwell in a community of their own making—a forest community—and unless I make an effort to penetrate that community, it can easily escape my attention.

Yet when I make the effort to embrace the tree, I feel its presence in a way that just looking at it or reading about it doesn't allow. The hug helps me understand that it's a living thing. As I spend more time with it, the network of relations becomes more apparent. It doesn't hug me back yet surely senses my presence. The human footprint is evident in more ways than any human will ever know.

I wouldn't say I love that tree, but I do value it as a member of my extended community. And I do have a relationship with the trees that surround my home. I will even go so far as to say they are intrinsic to my identity. They are my home. When I see them after a trip away, I feel somehow welcomed by the forest and woods that surround our cottage. I don't ask them whether they missed me. But I surely miss them.[16]

What do these trees teach me? How do I learn from them? I watch their flexibility in a driving rain. I notice how they serve as a home for hundreds of creatures. As Patricia, our botanist protagonist, reflects in *The Overstory*, even "a dead tree is an infinite hotel." Patricia views trees as "intricate, reciprocal nations of tied-together life that she has listened to all life long."[17] There's that word, or words, again. Reciprocal. Mutually corresponding. The return of favors. Tied-together life. Embodied networks. All our relations. Perceptual reciprocity. These collectivities of awareness that inform each other.

Please recall the ecological networks diagram reproduced earlier in this chapter as a metaphor for the entropic mind. Perhaps we can extend the metaphor even further to suggest that the same diagram portrays a network of reciprocal relationships—the extensive embodied networks of a forest community. Haskell in writing about trees as "nature's great connectors" is also referring to how a forest and indeed nature learn. This, too, is how we can strive to open our minds. Can we learn to perceive like a forest does, and how will that enhance our environmental learning? Can trees and forests also help us become good creatures?

Worlds within Worlds

The Earth contains a variety of different ecosystems: rainforests, grasslands, coral reefs, deserts, salt marshes, each with its own particular community of species. But a single animal is full of ecosystems too. Skin, mouth, guts, genitals, any organ that connects with the outside world: each has its own characteristic community of microbes. All of the concepts that ecologists use to describe the continental-scale ecosystems that we see through satellites also apply to ecosystems in our bodies that we peer at with microscopes.
—Ed Yong, *I Contain Multitudes*

An organelle inside an amoeba within the intestinal tract of a mammal in the forest on this planet lives in a world within many worlds. Each provides its own frame of reference and its own reality.
—Lynn Margulis and Dorion Sagan, *Microcosmos*

We are uncovering the multiple layers of perceptual reciprocity. We started with imaginary bubbles on a stroll through a vibrant meadow. The purpose of that journey was to better understand the unique perceptual worlds of different organisms. Then we considered how a deeper understanding of those perceptual worlds could promote interspecies communication and the various ways that could enhance our environmental learning. Next, using trees as an exemplar species, we expanded our view to entail the intricate connections that weave together a forest community, dynamic fabric of communications, many mediums of exchange, and multiple levels of embodied networks. And now we encounter the ubiquitous, invisible, microbial worlds that are the foundations of biological and ecological reality.

I find Yong's biogeographic metaphor an outstanding insight—one that is a fine mind opener. We can perceive the earth's biomes because they are visible at a human scale. With some basic ecological familiarity, we can distinguish the major biomes of our planet. We can distinguish biogeographic realms as well; we know the difference between a wetland and deciduous forest. In the previous section, Haskell explained that a close examination of a tree canopy reveals a surprising diversity of smaller-scale biogeographies. And then he reminds us how all these biogeographies and the species that reside within them form intricate networks of exchange.

But how often do we think of our own bodies as complex biogeographies? As Yong writes, "All of us have an abundant microscopic menagerie, collectively known as the *microbiota* or *microbiome*. They live on our surface, inside our bodies, and sometimes inside our very cells." The microbiota vary between organs, individuals, and species. He suggests that we think of ourselves as archipelagoes. "Each of our body parts has its own microbial fauna, just as the various Galapagos islands have their own special tortoises and finches."[18]

Our microbiomes vary in time and space, through the human life cycle. As a baby's diet changes, so do its gut microbes. Microbiomes also differ between human cultures and in different habitats. If this seems conceptually overwhelming, don't fret. Yong reminds us that "this is a rapidly changing field of science, and one still shrouded in uncertainty, inscrutability, and controversy." This is, Yong suggests, exactly what make the field so exciting. Yong offers an additional metaphor to enhance the biogeographic approach. He tells us that since early childhood, he has always loved visiting zoos. But on a recent visit to the San Diego Zoo, while writing his book, he realized for the first time that most of the life at the zoo is "invisible and inaudible." "At the main entrance, vessels full of microbes part with money so that they can file through gates and see differently shaped microbial vessels that loiter in cages and enclosures," he observes, adding that "trillions of microbes, hidden with feather-coated bodies, fly through aviaries. Other hordes swing through branches or scuttle through tunnels. One bacterial throng, nestled within the backside of a black hearthrug, fills the air with the redolent twang of popcorn. This is the living world as it actually is, and although it is still invisible to my eyes, I can finally see it."[19]

It turns out that the imaginary floating bubbles of umwelt contain multiple microworlds and are all communicating with each other. The more you study, the more complex it all gets, and the richness of the interactions surpasses what you can comprehend. The subtitle of Yong's book summarizes the learning experience of writing his book: *The Microbes within Us and a Grander View of Life*. By studying microbes and the role they play in the evolution of life on earth, in everyday biology, all aspects of human health, and ecosystem management, he opens his mind to a ubiquitous yet invisible reality. Most of all, it deepens his appreciation for the preciousness of life on the planet.

It's essential to recognize the exceptional educational work of Margulis and Sagan, who wrote close to a dozen books describing, portraying, theorizing, and philosophizing about microbiomes. Their accessible textbook, written in 1993, *Garden of Microbial Delights: A Practical Guide to the Subvisible World*, is a hands-on, illustrated approach to the microbial world. Margulis's work with Karlene V. Schwartz and Michael Dolan, *Diversity of Life: The Illustrated Guide to the Five Kingdoms*, is fine preparation for exploring Yong's biogeographic metaphor because as they visit dozens of familiar habitats and view the microbial worlds you'll find in each place. Margulis and Sagan also wrote *Microcosmos* and *What Is Life*, two highly recommended works that reflect on the ethical implications of the microbial evolutionary and ecological legacy. *What Is Life?* is one of the most mind-opening books I know for environmental learning. It frames an evocative narrative for the history of life on earth, expanding on the extraordinary transformations that accompany microbial evolution, as linked to the functions of fungal associations, photosynthesis, and the role of animals. The coauthors assert, "We have glimpsed ways of describing what life is: a material process that sifts and surfs over matter like a strange, slow wave: a planetary exuberance; a solar phenomenon—the astronomically local transmutation of Earth's air, water, and received sunlight into cells. Life can be seen as an intricate pattern of growth and death, dispatch and retrenchment, transformation and decay."[20]

A better understanding of the microbial world expands how we think about the human presence on the planet. It encourages respect and humility for the interconnectedness of all living things, the biogeochemical cycles, and the embodied networks, both visible and invisible, that contribute to human flourishing in a dynamic biosphere. Such understanding is the foundation for an encompassing perceptual reciprocity. We are, as Yong underscores, living islands, containing invisible microworlds, in a sea of billions of other beings, differentiated by skins and faces that may seem like separate selves, but are all swimming together in the ocean of life.

Memory Unfolds in Places

Places remember what people forget.
—Richard Powers, *The Overstory*

Perhaps the greatest human intellectual accomplishment is the ability to decode the information residing in planetary memory. Consider all the remarkable ways this occurs. By reading ancient rocks and minerals, studying microscopic fossils in relic environments, understanding radioactive decay and isotopic analysis, and decoding the intricate strands of DNA, and through many other measures, humans read the earth's memory to figure out the age of the planet. We can now (within ever-finer estimates) re-create the geologic, atmospheric, biological, and biogeochemical history of the earth. With the naked eye and on a more limited time scale, let's say from ten to ten thousand years, you can read the memory of your local landscape. Where I live, it can be found in glacial remnants, stone walls, beaver dams, tree rings, soil series, and aboriginal relics. If I know where and how to look, I can read the memory of the forested landscape.

In his fine book *Mannahatta* (previously introduced in chapter 7), Sanderson describes the techniques his research team used to create a portrait of the natural history of Manhattan in September 1609. Their methods are a combination of understanding who lived on the landscape at that time (the Lenape tribe), the various species that coinhabited the place and the communities they formed—how "the forests, streams, wetlands, and beaches . . . preceded TriBeCa, Chelsea, Midtown, and Harlem," and how those communities formed "a dense network of ecological relationships," or what they characterize as Muir webs. They also used old maps to trace the human footprint on the landscape and its ecological consequences. Their goal was not to return Manhattan to a primeval state but rather to "discover something new about a place we all know so well" and find the "green heart" of New York City compelling a "new start to history."[21] Sanderson's team aspires to reenvision Manhattan by resurrecting the memory of the recent past, reminding us that only four hundred years ago, there were fifty-five distinct ecological communities on Manhattan Island—a remarkable diversity of landscapes and species, or what Sanderson describes as ecological neighborhoods.

These so-called Muir webs present another way to think about memory. Haskell suggests that the forest community represents a multispecies symbiotic association of information transmission, including forms of collective memory. "Roots and twigs have memories of light, gravity, heat, and minerals." Chickadees place caches of balsam fir seeds in rotting logs and other seedbeds throughout the forest. "Bird memories are therefore a tree's dream of the future."[22]

We tend to think of memory as reflected in human experience, as what is stored in our DNA, minds, and bodies, as what is written in history books, or transmitted informally among friends and neighbors. Yet we know all too well that memory is frequently contested, and whoever controls the pathways of information will manipulate memory to tell their own version of that history. It's challenging enough to clarify, sharpen, and recollect our own experiences accurately, as our impressions are so easily transformed by what we imagine to be true or dream. When we share our memories with others, we often realize that we had different impressions of the same moment in time. How easily that leads to miscommunication or disagreement. How often, too, we live in the past to ignore the present, relying on recollections that comfort us in the face of uncertainty. Similarly, we bond with people when we share similar recollections, and our collective agreement may take on a life of its own.

By extending our concept of memory to include the geologic record, landscape, and multispecies symbiotic associations (Muir webs)—all the environmental changes that contribute to our ecological neighborhood—we expand our perspective beyond human memory. We enhance our understanding of the place where we live. When people refer to "the spirit of place," I believe this is what they mean: a collective ecological memory that connects people, species, and landscapes. That spirit may be embodied in a landscape feature, perhaps a mountain or lake, or a human artifact like a shrine. Think of the Statue of Liberty in New York Harbor, not only as a beacon of freedom, but as a shrine on a gateway to an extraordinarily diverse grouping of ecological and cultural communities. What makes any place special is that it provides a focus for collective ecological and cultural memory. What makes it special is how it also serves to open our minds.

Memory is a reciprocal process. It extends way beyond human experience. The earth's history informs the landscape. The landscape informs the creatures who dwell within it. The creatures inform the microbes within themselves. And the microbes shape the history of the earth. Memory unfolds in multiple ways. To perceive environmental change, we must be aware of multiple memory pathways because they each provide a glimpse into our connections with the world beyond ourselves.

Perceptual Reciprocity in the Neighborhood

It thrills her to sit at meals and be part of the laughter and shared data, the dizzy network trading in discoveries. The whole group of them, looking. Birders, geologists, microbiologists, ecologists, evolutionary zoologists, soil experts, high priests of water. Each of them know innumerable minute, local truths. Some work on projects designed to run for two hundred years or more. Some are straight out of Ovid, humans on their way to turning into greener things. Together, they form one great symbiotic association, like the ones they study.
—Richard Powers, *The Overstory*

You missed that. Right now, you are missing the vast majority of what is happening around you. You are missing the events unfolding in your body, in the distance, and right in front of you.
—Alexandra Horowitz, *On Looking*

Let's return for now to the more appropriately familiar scale of human interactions with the landscape, the places where we dwell, work, and play, the weather and sky, the buildings and fields, the lakes and rivers, the human community, and the places we call home. These places are the ground floor for how we come to know the world. And by paying close attention to them, we can use the familiar as a guide to the unknown. Perceptual reciprocity begins in your neighborhood.

Every nook and cranny of a neighborhood is a treasure chest of collective memories if you know how and are willing to look. Horowitz's book *On Looking* is a wonderful resource for neighborhood exploration, presenting multiple views of a New York City street, and the approach she takes has unlimited applications in educational settings or simply to expand how you think about what you've come to take for granted. She describes herself as a "sleepwalker on the sidewalk," and her goal in writing her book is to "knock herself awake." Horowitz sets out to take a series of walks with a variety of "experts," or at least individuals who see things differently than she does. She takes her first walk with her two-year-old son and tries to enter his world as best she can (spoiler alert: it really tests her patience!). She takes her final walk with her dog, aspiring to enter her umwelt of olfactory stimulation. For the other walks, she invites scientists and scholars to join her. Horowitz organizes these walks around three sequential themes—the inanimate, animate, and sensory city—and her companions include a geologist, wildlife biologist, naturalist, artist, urban geographer, medical doctor, blind social worker, and sound designer. The geologist, for example, points out how a "random sample of any block in the city or any city" contains "the history of geology across eras and locales."[23]

She chooses her companions based on her friendships, experiences, and interests. You might live on the same block and choose an entirely different crew. How interesting it would be to take that same walk with people from different ethnic backgrounds, recent immigrants, and people who come from an entirely different neighborhood and are walking the block for the first time. Or to enter the umwelt of floating bubbles to explore the neighborhood from the perspective of a squirrel, pigeon, maple tree, or migrating warbler. Or you can walk around the block using various technological aids—binoculars, a hand lens, kaleidoscope, camera, or recording device—and notice how your perspective changes in each case.

Powers reminds us above that environmental learning can and must be a "network trading in discoveries," and people who engage in that network also take great pleasure in a thriving community of environmental learners. They study what's close at hand, knowing that their research may require several hundred years of data. His menagerie of traders reflects the spirit of the global long-term environmental change research community (see chapter 8).

Such initiatives need not be merely the province of experts. Any community can take this on. Many are already doing so. Review the work of the Project for Public Spaces to see how urban communities use "placemaking" to engage neighborhoods in strategic planning based on a collaborative vision of place.[24] Or check out the India-based Barefoot College, which works with rural communities to build collaborative, sustainable, antipoverty coalitions with an emphasis on leadership, education, and community transformation.[25]

Perceptual reciprocity is a community-based effort. It requires diverse human perspectives of expertise and experience as well as the ability and willingness to tap into the multiple ways of knowing intrinsic to the biosphere—from umwelts to networks to microbes. Perceptual reciprocity shares the collective memory of a place—human stories, fossil relics, tree rings, landforms, and shrines. Perceptual reciprocity engages short-term memories of what happened yesterday with the long-term memories of the movements of peoples, species, and continents. In all these ways, perceptual reciprocity culminates in collective environmental learning, the best formula we have for ensuring thriving communities in flourishing ecosystems, and a recipe for building an ecological neighborhood.

Generosity and Environmental Learning

Please refer again to the evocative diagram (figure 9.1) that portrays both how the mind shifts during a psychedelic experience and the complexity of an ecological network. This diagrammatic portrait is merely a two-dimensional metaphor for what happens when you open your mind. Throughout this chapter and entire book, I suggest that the purpose of environmental learning is to better understand how to open your mind—looking more deeply into the dynamic relationships of place and planet, and they are informed by the ecological and evolutionary matrices of our lives. How do we manifest that understanding? How do we learn to explore it? How does it help us confront the challenging issues of our times while also enriching our lives?

Opening our minds is always a melding of self and other, identity and community, tribe and territory, organism and environment, and the ordinary and ineffable. It occurs across the vast social and ecological networks of place and space, and through the evolutionary and cultural histories of time and generations.

It is not easy to do! The theory of the entropic mind explains why. When there are too many connections, the mind lacks focus, it's directionless, and it may never find its way home. That's when you need Harold's purple crayon! You can only take in so much at one time. When the connections are too few, then the mind retreats into the illusion of safety and control. You fall prey to recursive loops, repetition, and stagnation. The educational challenge is how to expand awareness strategically, developmentally, and compassionately. That's why we need mentors and teachers—people and communities that provide guidance along the way. Environmental learning and the expanded awareness it entails demands the finest teachers and mentors.

It also requires reciprocity—the ability to share, collaborate, find middle ground, and expand awareness by comparing what you perceive as well as value to other perspectives. You need not relinquish what you believe. Rather, use your belief system as an improvisational guide to what may come next. Who knows? Maybe your belief system will subtly shift. And maybe your perceptual world will find new frontiers of learning.

What will all this eventually yield? What happens when you find the middle ground, blending perceptions along with the shared perspectives of multiple peoples, generations, cultures, networks, creatures, and landscapes? Under the best circumstances, a truly reciprocal encounter yields generosity.

Reciprocity implies exchange. Generosity implies gratitude, kindness, and compassion. It encourages empathy, dialogue, connectedness, and love—the giving of yourself to others. Reciprocity is the sharing of perspectives and ideas. Generosity is sharing what is most precious to you and doing so with no exchange in mind. It is a gift of kindness and love.

In my experience, the most inspiring life moments are encounters with the generosity of others. How wonderful it is to be around generous people! They have so much to teach us. Their gifts of kindness and compassion stimulate similar responses in ourselves. They fill us with hope and love.

Is it too outlandish to suggest that the biosphere is an expression of planetary generosity? Four and a half billion years of biospheric processes yields a magnificently biodiverse ecosystem with hundreds, and even thousands, of habitats that supply homes and nourishment for countless species, including ourselves. Our very ability to read and write these words is a gift of monumental proportions. Our very ability to potentially expand our perceptual awareness is utterly splendid and perhaps unsurpassed on this planet. The vast unknowns that surround us only deepen the mystery for which we should be forever grateful. What is it that we can offer in return for these gifts? What might we give back? What is the best expression of our generosity? These questions, too, are the essence of environmental learning and foundation of how we come to know the world.

Perhaps the best way to enhance environmental learning is to spend time with the extraordinary people who devote their lives to these expressions of generosity. You will find them everywhere—in cities and the most remote countrysides, in the diverse habitats and ecosystems of the earth—representing every conceivable age group and ethnicity. They are working on sustainable solutions, antipoverty strategies, urban ecologies, human and ecosystem health, intercultural understanding, peace and justice, gender equality, LGBTQ issues, animal welfare, and wilderness preservation. You could make a list of good causes and organizations as long as this book. And you could enter their names in a field guide to human generosity that will entail volumes more. These are the heroes of our times, and they are working together, against formidable challenges, to understand the tides of change, and promote solutions filled with kindness, compassion, empowerment, and generosity. They are the contemporary capstone of planetary environmental learning. They live in your neighborhood. They work on these projects every day. They have been doing so for generations. And they are most probably you.

A Field Guide to Recommended Readings

One of the great pleasures of writing a book is the opportunity to read widely, dwell in the world of ideas, and develop an intellectual sense of place. Yet we live in a time of unprecedented intellectual scope. Indeed, it can be overwhelming to keep up with all the books and publications that are pertinent to your interests. You do the best you can to search for what you think you need. I find it helpful to include a healthy dose of browsing. Take a serendipitous journey through an interesting bookstore or library, and roam through sections both familiar and unfamiliar. Who knows what you will find?

This list of recommended readings isn't close to being comprehensive. Consider it a brief field guide to some of the subjects covered in *To Know the World*. These are works I find particularly interesting and rewarding. I mainly include books that are timeless and important. They are essential to an informed discussion. They will be of interest whenever you check them out. Works that don't appear here but are listed in the bibliography are well worth your attention too. This list is a way to get started. For organizational efficiency, I follow the chapter sequence, but most of these books are relevant for many of the ideas I cover. I don't include periodicals, reports, and websites, so please use the footnotes and bibliography accordingly. And for the full titles of the books listed here, please see the bibliography.

The Past and Future of Environmental Learning

A wonderful way to think about environmental learning is to read biographies of exemplary thinkers. I especially recommend reading about the nineteenth-century pioneers. Laura Dassow Walls's biography of Thoreau, Andrea Wulf's biography of Humboldt, and Janet Browne's two-volume biography of Darwin are all outstanding. In the twentieth century, Rachel Carson is a seminal environmental thinker. See the biographies by Linda Lear and William Souder. Mark D. Hersey's environmental biography of George Washington Carver presents a remarkable early twentieth-century African American vision for sustainable agriculture—a crucial dimension of environmental learning.

For a multicultural perspective on late twentieth-century environmental issues, with an emphasis on cultural identity, please see the excellent anthology by Alison H. Deming and Lauret E. Savoy, *Colors of Nature*.

Gary Snyder was one of the most influential environmental thinkers of the late twentieth century. For a more comprehensive look at this poet's work, see *The Gary Snyder Reader*.

Memory Forever Unfolding

I first encountered Lewis Hyde's work in the 1970s when I used *The Gift* as a way to place trade, exchange, and reciprocity in an ecological perspective. His pithy and wise book *A Primer for Forgetting* brilliantly weaves memoir through personal awareness, Greek mythology, the US civil rights movement, dangers of nationalism, and reconciliation. Hyde's work is filled with meaningful "developmental interludes." Many autobiographical works use memoir as a means to illuminate environmental learning. Four of my favorites are Hope Jahren's *Lab Girl*, Robin Wall Kimmerer's *Braiding Sweetgrass*, Sy Montgomery's *How to Be a Good Creature*, and Lauret E. Savoy's *Trace*. Savoy's book is a stunning meditation on place, identity, geology, and race.

For another illuminating approach—the relationship between time, memory, and physics—see Carlo Rovelli's *The Order of Time*. And for a comprehensive overview of the psychology of awareness (memory always unfolds deep in the layers of the psyche), see Evan Thompson's *Waking, Dreaming, Being*.

The Tides of Change

A good way to better understand the intricate connections between the tides of change (environment, equity, diversity, and democracy) is to read the latest literature on climate justice. Three of the stalwart, prodigious writers and activists are Bill McKibben (*Falter*), Naomi Klein (*On Fire*), and Mary Robinson (*Climate Justice*).

To look more deeply into the timeless, historical challenge pervading these questions, especially regarding tribalism and inequality, I find three books indispensable: Robert Sapolsky's *Behave*, Mark Pagel's *Wired for Culture*, and Kent Flannery and Joyce Marcus's *The Creation of Inequality*. Robert D. Bullard's *Unequal Protection*, Julian Agyeman's *Introducing Just Sustainabilities*, and Dorceta E. Taylor's *The Environment and the People in American Cities* are academic pioneers of the environmental justice movement, and all of these writers have a significant body of important work. For an excellent overview of the North American indigenous perspective on environmental justice, see Dina Gilio-Whitaker's *As Long as Grass Grows*. Rob Nixon's *Slow Violence and the Environmentalism of the Poor* provides a literary and political perspective on the foundations of environmental justice.

Is the Anthropocene Blowing Your Mind?

To better understand the ecological dynamics of the changing biosphere, check out landmark studies. Then you can decide for yourself whether the Anthropocene concept is hype or reality. My favorite is the technical and accessible anthology by W. Steffen (and ten other contributors), *Global Change and the Earth System*. For comprehensive coverage of the biosphere concept, see the prodigious work of Vaclav Smil, especially *The Earth's Biosphere*, and then read the original conceptualization by Vladimir Vernardsky, *The Biosphere*. Tyler Volk covers the metabolism of earth's biogeochemical cycles in *Gaia's Body*. For a challenging but important work, see Tim Lenton and Andrew Watson's *Revolutions That Made the Earth*.

For a recent historical perspective, see Rosalind Williams's *The Triumph of Human Empire*, which traces the life and times of Jules Verne, William Morris, and Robert Louis Stevenson to interpret how humans came to seemingly dominate the globe in the nineteenth century.

Amitav Ghosh's *The Great Derangement* is an outstanding book that exposes the hubris of European expansion, colonization, and how a bourgeois mentality promotes climate denial. The book is beautifully written and filled with evocative wisdom. Frank Trentmann's magisterial history of consumption, *Empire of Things*, explores the universality of materialism and how it changes how we perceive the world

If you're really ambitious and want to dive into the existential abyss of artificial intelligence, check out Max Tegmark's *Life 3.0*. Yuval Noah Harari's *Homo Deus: Brief History of Tomorrow* covers many of the themes in this chapter. Matthew Crawford's *The World beyond Your Head* has a great deal to say about clarifying intention and resisting attention competition.

Constructive Connectivity

Several recent historical works focus on human networks. Niall Ferguson's *The Square and the Tower* contrasts network organization and hierarchies, mainly in European history. Peter Frankopan's *The Silk Roads* is a fascinating historical account of Eurasian trade networks. In *Connectography*, Parag Khanna describes how various global networks are emerging and evolving. Sociologist Charles Kadushin clearly introduces the social science of network analysis in *Understanding Social Networks*.

To pursue biosphere network archetypes, start with the introductory chapters of Paul Stamets's *Mycelium Running*, where he describes mycelial networks as the original internet. David George Haskell's enriching *The Songs of Trees* illuminates the intricate connections in forests.

To peruse the extraordinary array of spectacularly illustrated books on network visualization, especially as pertaining to ecology, the biosphere, and globalization, start with Manuel Lima's *Visual Complexity*. But don't neglect Sarah Rendgen and Julius Wiedemann's *Understanding the World* or Alistair Bonnett's *An Uncommon Atlas*.

Katharine Harmon's *The Map as Art* can be read for its aesthetic wonders, but also as a compendium of teaching ideas for visualizing networks. Steven Johnson's *Where Great Ideas Come From* provides some interesting approaches to constructive connectivity, both to better understand creativity and inspire curricular approaches.

Migration: The Movement of People and Species

The migration of people and species in the biosphere is both fascinating and instructive. James Cheshire and Oliver Uberti's *Where the Animals Go*, featuring beautiful maps and well-written essays, illustrates wildlife movements around the world. David Wilcove's *No Way Home* is a fine survey of ecological and climatic threats to animal migration. Geerat J. Vermeij's *Nature: An Economic History* is dense, but it provides excellent examples of species migration from an adaptational perspective over five million years. Ron Redfern's *Origins* neatly displays the movement of oceans and continents, and how that impacts species migrations. For a popular approach to bird migration and contemporary threats to it, see Kenn Kaufman's *A Season on the Wind*.

The literature on the impact of climate on human migration is proliferating. For two academic anthologies, see S. Irudaya Rajan and R. B. Bhagat's *Climate Change, Vulnerability, and Migration*, and Simon

Behrmann and Avidan Kent's *Climate Refugees*. For a powerful progressive approach, see Todd Miller's *Storming the Wall*. And the photographic collection *Climate Refugees* by Collectif Argos is stunning.

Two recent novels that bring a human dimension to migration are Moshin Hamid's *Exit West* and Amitav Ghosh's *Gun Island*. Anthropologist Anna Lowenhaupt Tsing's *The Mushroom at the End of the World* looks at uncertainty as a precondition for survival in the Anthropocene. It's a brilliant treatment of life experience among specific groups of migrants.

Adam Rutherford's *A Brief History of Everyone Who Ever Lived* and David Reich's *Who We Are and How We Got Here* both provide excellent coverage of the latest human DNA research, especially as it pertains to the movement of peoples and genetic ancestry.

For some additional ideas regarding mapping migration in your community, see Katharine Harmon, *You are Here: NYC: Mapping the Soul of the City*, and Rebecca Solnit and Joshua Jelly-Schapiro's *Nonstop Metropolis: A New York City Atlas*.

Cosmopolitan Bioregionalism

Gary Snyder's wonderful book of essays *The Practice of the Wild* neatly discusses the bioregional idea while applying it to daily life. See especially his classic piece "The Place, the Region, and the Commons." Also see the classic academic anthology edited by Michael Vincent McGinnis, *Bioregionalism*. Tom Lynch, Cheryl Glotfelty, and Karla Armbruster's anthology *The Bioregional Imagination* contains many excellent essays, and is particularly relevant from both a literary and educational perspective. For some of the challenges inherent in local/global relationships and their implications for citizenship, see Ursula Heise's *Sense of Place and Sense of Planet*. For a superb multicultural and multigenerational anthology on community, identity, and place, see Annick Smith and Susan O'Connor's *Hearth*.

Mark Dowie's book *The Haida Gwai Lesson* is a superb discussion of indigenous issues, while offering "lessons" in bioregional and tribal autonomy within a larger state system. For multiple native voices on place, displacement, and identity, see the poetry anthology edited by Heid E. Erdrich, *New Poets of Native Nations*.

Eric W. Sanderson's groundbreaking book *Mannahatta* is a must read if you are interested in thinking about urban ecology, urban history, and sense of place. Marina Alberti's *Cities That Think like Planets* is a comprehensive synthesis of urban ecology insights. Paul Knox's *Atlas of Cities* brings a useful historical perspective. Doug Saunders's *Arrival City* looks at the migration between urban and rural settings. Benjamin Barber's *If Mayors Ruled the World* and *Cool Cities* both suggest that cities are the only political jurisdictions that can actually accomplish ecologically oriented policy objectives. Browse through the sumptuous anthology by Michael Mostafavi and Gareth Doherty, *Ecological Urbanism*, for a portfolio of urban, bioregional, and sustainable solutions.

I highly recommend Robin Cohen and Olivia Sheringham's *Encountering Difference* for a deeper understanding of the challenges and potentials of an inclusive cosmopolitan vision. Also see Kwame Anthony Appiah's personal and philosophical work *Cosmopolitanism*. Helen Thorpe's *The Newcomers* is a well-written case study of an ongoing challenge: how high schools work with migrant students. The book shows what's possible when committed teachers work together in a community.

Improvisational Excellence

It's a stretch to claim that the science of evolutionary ecology describes improvisational processes in nature. Nevertheless, you can interpret adaptational response as a form of creative improvisation. In that regard, I suggest Geerat J. Vermeij's *The Evolutionary World*.

For more on pattern-based environmental learning, I recommend the technical yet accessible text by Timothy F. H. Allen and Thomas W. Hoekstra, *Toward a Unified Ecology*. To study and take pleasure in the aesthetic dimension of patterns in nature, check out Phil Ball's *The Self-Made Tapestry* and *Patterns in Nature*. Tom Wessels's *Reading the Forested Landscape* teaches you how to observe patterns in the forest, and Tristan Golley's *How to Read Water* explores patterns in water, as does Theodor Schwenk's *Sensitive Chaos*.

Robert Macfarlane's remarkable book *Landmarks* examines the wonderful diversity of words that describe patterns, both ephemeral and residual. And David Lukas's *Language Making Nature* is filled with great ideas for inventing your own words. Both books are must reads for anyone who teaches and thinks about environmental learning.

There are musicians who think about patterns in nature, their correspondence with music, and the origins of improvisation. The master of these correspondences is David Rothenberg. Start with *Sudden Music*, follow it with *Survival of the Beautiful*, and then read any of his three books *Why Birds Sing*, *Thousand Mile Song*, or *Bug Music*. Bernie

Krause is a pioneer in acoustic ecology. See his *Wild Soundscapes* or *The Great Animal Orchestra*. Stephen Nachmanovitch's excellent book *Free Play* is filled with wisdom about creative improvisation. For a more academic approach to imagination and improvisation, also by a musician, see Stephen T. Asma's *The Evolution of Imagination*. Anyone who listens to or plays music, and is interested in improvisation, must read W. A. Mathieu's *Bridge of Waves*.

My favorite translations of the *I Ching* are comprehensive, interpretative, and accompanied with lengthy as well as helpful introductory commentary. I recommend John Minford's *I Ching*, Jack Balkin's *The Laws of Change*, and Deng Ming-Dao's *The Living I Ching*.

Perceptual Reciprocity

Michael Pollan's *How to Change Your Mind* is ostensibly about the history and promise of psychedelics, but the book is relevant to anyone who is interested in expanding personal awareness. David Abram's *Becoming Animal* is about exploring and widening perceptual awareness as well as sensitivity. Also see Andrea Weber's *The Biology of Wonder*. Alexandra Horowitz's *On Looking* is a perspective changer, as is Jakob von Uexküll's *A Foray into the World of Animals and Humans*.

I highly recommend Richard Powers great novel *The Overstory*. It's both a testimony to how humans develop relationships with trees and fascinating history of Americans from all walks of life who develop those relationships. Eduardo Kohn presents an anthropological view of *How Forests Think*. Carl Safina's *Beyond Words* and Alison H. Deming's *Zoologies* promote animal respect and kinship, while Peter Godfrey-Smith's *Other Minds* is about the octopus. Caspar Henderson's *The Book of Barely Imagined Beings* is the most respectful bestiary I've ever read. Ed Yong's *I Contain Multitudes* opens you to kinship with microorganisms. Lynn Margulis and Dorion Sagan's *What Is Life* bring an evolutionary ecology perspective to all the organisms that dwell on this planet. This book, I promise, will change your life!

Words That Matter: A Glossary

To Know the World is filled with unique word combinations as well as some reconsiderations of familiar words. This isn't a glossary of formal definitions but rather an interactive index where you can think about an expression and consider its meaning. Each of these words also appears in the standard index so you can see how it's used in the narrative. In the spirit of reflective learning, I've attached a question (in italics) for each word, designed to stimulate further conversation or curricular application.

Anthropocene Essentially the Anthropocene concept suggests that the contemporary era qualifies as a new geologic time frame, signifying that the human practice of natural resource extraction is now a geologic force dramatically impacting the biosphere. *When does the Anthropocene begin, and what are its ethical implications?*

Autonomy Autonomy is the capacity of individuals to make decisions, take actions, and manifest behaviors on their own accord. *Is autonomy an aspiration or a virtue?*

Biosphere network archtypes Network patterns are a biosphere phenomenon at multiple scales—from mycelia to species migrations, from shipping routes to the internet. I suggest they are archetypal forms, prevalent throughout the planet and perhaps the universe. *What function do these archetypes serve, and how do they help observers connect social and ecological processes?*

Constructive connectivity Connectivity is constructive when it promotes creativity and innovation, facilitates communication between diverse cultural communities, and illuminates the relationship between ecological and social networks. *What are the connections that matter in building social capital, community trust, ecological understanding, and human flourishing?*

Contact zones Contact zones are the geographic and cultural settings where multigenerational as well as multicultural communities have planned or serendipitous encounters. *How can contact zones promote constructive connectivity?*

Cosmopolitan bioregionalism Bioregionalism is the concept that geographic, cultural, and ecological criteria should be the foundation of local political economy. It emphasizes place-based community relationships. Adding cosmopolitan broadens the concept to recognize that local communities must also strive to achieve multicultural diversity, understand the relationship between local aspirations and global dynamics, and find ways to meld urban and rural, local and global, and community and planet. *How do we expand the concept of local place to include biosphere processes?*

Curate Curating is the process of gathering and collecting ideas, interpreting them according to specific criteria, and organizing those ideas to promote access and discourse. Curation, once considered the specific role of museums, is now necessary with any information-oriented project, from websites to libraries, from public art to twitter. *What is the role of curators in facilitating environmental learning?*

Deliberate pause This involves taking the time to slow down, reflect, and redirect one's gaze to biosphere processes along with the more-than-human world. *How do we promote deliberate pauses in all aspects of our lives?*

Developmental interludes Such interludes build narratives based on seminal life moments—events and memories that had a profound impact, and represent (in retrospect) learning watersheds over the course of a lifetime. *What is the role of developmental interludes in constructing collective narratives of environmental learning?*

Ecological identity The reason for melding these words is to suggest that how we perceive ourselves in the world (identity) can be integrated with our ecological surroundings. We project personal identity beyond the self, and see ourselves as an organism in context with other species, habitats, and the biosphere. *How does internalizing an ecological identity change how you perceive education, community participation, and environmental activism?*

Ecosystem services I link these words to reiterate that ecosystems and the biosphere are the fundamental substrates of human flourishing, and promote a deeper understanding of how we perceive so-called natural resources, stretching that understanding beyond cost, benefit, and profit. *Take a walk and try to identify how the immediate infrastructures of your life are manifestations of ecosystem services.*

Environmental learning Environmental learning entails the process of expanding awareness about the biosphere, deepening understanding of the human relationship with species and ecosystems, and promoting ideas and behaviors that apply those insights to all aspects of daily life. *How can environmental learning become the foundation of education, community participation, and civic discourse?*

Filter Filtering involves creating a boundary that only allows certain materials, ideas, and concepts to pass through. Such boundaries may prevent threats to a system (or individual), but they also may be restrictive by preventing vital information to pass. *What is the role of filters in both opening and closing minds?*

Fluid sovereignties An example of applied bioregionalism, fluid sovereignty is the idea that a community (or nation) may be a participant in multiple civic principalities, sharing autonomy and negotiating authority depending on changing environmental circumstances. *Do fluid sovereignties promote robust climate solutions?*

Golden spike The golden spike is the nail that fastened the final railroad ties connecting the US East and West, thus completing the transcontinental railroad. The expression also signifies the watershed event(s) that represent the onset of the Anthropocene. *Why is the spike referred to as golden, and what does that signify?*

Homophily This is the tendency of people to congregate (in networks) around other people with similar interests, values, and backgrounds. *Why is it so difficult to break out of homophilic networks?*

Improvisational excellence Improvisational excellence is the aspiration to cultivate creativity, insight, and flow experiences in rapidly changing learning settings, as exemplified by music, sports, dance, comedy, organizational leadership, and observing the biosphere. *Why is improvisational excellence a powerful approach for navigating the tides of change?*

Liquid networks This involves the ability to move freely between multiple networks and ideas, thereby stimulating creativity and innovation. Steven Johnson describes carbon as the original connector as it is capable of hosting so many different elements and is a facilitator for the origins of life on earth. *What are some examples of liquid networks in ecosystems?*

Mindfulness Derived from the Buddhist meditative tradition, mindfulness is the process of clearing your mind by focusing your awareness on the present moment. *How does mindfulness contribute to environmental learning?*

Network A network is a series of intricate pathways that facilitates the flow of information, energy, and matter, manifesting in topological patterns and forms that transcend scale. *What is the relationship between networks and hierarchies?*

Pattern-based environmental learning This is the process of observing and interpreting patterns in the biosphere, including cycles, networks, fractals, flows, and boundaries, among many others, to enhance an understanding of environmental change. *Is there an appropriate developmental sequence for teaching pattern-based environmental learning?*

Perceptual reciprocity Every organism on the earth perceives. Collectivities of perception emerge when organisms adapt and coevolve in changing environments. Communities of species (forests, habitats, and even landscapes and the biosphere) perceive. Collectivities of awareness are mutually responsive in multiple scales of place and time. Apprehending environmental change requires a deepening of this awareness. *What is the difference between perceptual reciprocity and sentience?*

Portal A portal is a place (virtual or visceral) that transports you to another place or opens the gates to infinite varieties of new conceptual places to explore. *Which portals do you choose to open, and which do you keep closed?*

Proliferation This is the process by which any idea or concept can rapidly reproduce into infinite branches of possibility. *How might proliferation be curated to enhance environmental learning?*

Ubiquitous novelty Ubiquitous novelty is the expectation that every new experience is unique and unprecedented, as opposed to the expectation that any one experience has great depth and meaning. *When is novelty a burden, and when does it open new portals?*

Environmental Learning Templates

To Know the World covers twelve environmental learning activities—templates for expanding perception, observation, and interpretation. Each template emphasizes the importance of enhancing the imagination, facilitating improvisation, and building substantive skills and experience. "Template" refers to frameworks: patterns of thinking and instruction that initiate fulfilling conversation processes. These activities involve changing perspectives, transcending scale, and promoting mindfulness. They are educational processes that cultivate insights for environmental learning. Please use them in creative ways, modify them accordingly, and change them based on your interests, and then drop me a line to let me know what worked best and if you came up with a wonderful new approach. Similar to the "Words That Matter" section, please use this as an annotated index and refer back to the text for the details.

Portals: Screens, Windows, and Frames

What happens when you switch perspectives between screens (especially cell phones, tablets, and laptops) and the wide field of your visual as well as acoustic surroundings? How is your perspective narrowed, circumscribed, filtered, or focused by the use of screens? A good introduction to this suite of potential activities is to consider the various ways that visual boundaries frame our perspective—from window to doors, from photographs to bookshelves—and how our perspective shifts when we move between frames. When these milieus change, how is our perception of the natural world altered? You can do this acoustically too with the use of earbuds or by experimenting with various soundscape environments.

Reading the Day: Passing Moments of Environmental Learning

With the popularization of yoga, meditation, and other mindfulness techniques, there is a growing familiarity with reflective practices crunched into short time frames—something of a contradiction, but responsive to the pace of most people's lives. Why not consider "five-minute" moments of environmental learning, when you place all your attention on the ecological surroundings? The key here is to find the "passing moments" during the course of a day—moving from indoors to outdoors, stepping out of a vehicle, walking between destinations, opening the window and gazing at the sky, and so on.

Mapping Migration

As migration becomes the "perfect storm" within the tides of change, it will impact everyone's community. You can internalize the migration experience by tracing your family history. How did you arrive where you are? Who preceded you, and what path did they travel? Mapping these journeys demonstrates the mobility of just about every human who lives on this planet. Juxtaposing human migration with other species is equally revealing. Comparing these maps in group settings vividly illustrates migration as a biosphere phenomenon and builds solidarity between people of diverse backgrounds. Please refer to the curricular sequence in the text: migration and the biosphere, migration in your human community, and migration and you.

Maps and Stories

Mapping activities are community equalizers, allowing people with diverse skills to reflect on the circumstances of their lives. I've used personalized, narrative maps as teaching tools for my entire career and am continually inspired by the imaginative possibilities. The text suggests a seven-step sequence of activities that are easily modified depending on the community setting: what is indigenous, travelers and newcomers, mapping diversity, contact zones, encountering difference, connecting local and global, and ecocosmopolitan sense of place.

Observing Networks

A wonderful outcome of this observational activity is how it illuminates what we so often take for granted: the connections that are the foundation of ecological infrastructure and built environments. For example, the next time you travel somewhere, focus on all the wires in the landscape. Where do they come from, and where do they go? What would a landscape without wires look like? When did wires first appear? Wires carrying electricity to my home allow me to write this passage, but as I look out the window, I don't think they are an attractive landscape feature.

Navigating Networks

Once you identify networks (at multiple scales), you can think carefully about how to curate them. Such a thought process has both a personal and organizational impact. What networks do you belong to? What networks influence your decisions? Is this influence direct or subtle, or perhaps totally outside the realm of how you think you make decisions?

Sense-of-Place Map

For almost four decades, I have used the sense-of-place map as an educational rite of passage. Mapmakers get to know each other based on their place affiliations, interests and values, and ecological identities. Also, they recognize the wisdom that emerges when people share life narratives, developmental interludes, and personal histories within the context of place, habitat, and ecosystem.

Language Making Nature

Wonderful insights emerge when you deconstruct words, find their origins, and then develop unique word combinations based on pattern-based environmental learning. Start with the etymology of words you think you understand. Consider the linguistic, cultural, historical, and then ecological meaning of the word. How has the meaning changed? Then after a walk in the woods or stroll down the streets of your city, invent words that describe patterns there may not be a word for.

Pattern-Based Environmental Learning

Frequently natural history observation emphasizes species identification, prey-predator relationships, food chains, and basic ecological concepts. A pattern-based approach prepares you to interpret environmental change. It also allows you to ascertain patterns even if you cannot identify species (as important as that may be). I'll briefly describe a versatile teaching activity: "observing what you observe." Ask a group to take a fifteen-minute walk and instruct the participants to note every observation. Suggest their notes may include typical natural history observations, but also patterns such as waves, flows, cycles, topologies, or whatever else comes to mind. When the group reconvenes, have the participants share their observations and then discuss what patterns are most illuminating in perceiving environmental change.

I Ching and Environmental Change

Take the metaphoric representations of the eight trigrams—Heaven, Wind, Water, Mountain, Earth, Thunder, Fire, and Lake—and rearrange them in all the permutations, sixty-four combinations in all, representing a field guide to environmental change processes. Read the day (see above) by considering which combination captures what you perceive. How does that combination reflect your place in the biosphere, state of mind, and immediate prospects for the day? I believe the *I Ching* has great potential as a biospheric knowledge system, linked to human wisdom and practical ethics, and should be more widely known and used for this potential.

Floating Bubbles

The challenge is how to change perspectives so as to enhance perceptual capabilities. Imagine yourself as another species (what does the red-tailed hawk see?) or enter an inaccessible sensory framework (infrared, ultraviolet, echolocation, or magnetic waves). This can be conceptually expansive, especially when supplemented with good research and reading. Why restrict this to organisms? Flow like the stream, ride the wind, become a carbon molecule, or trace your history as a polished stone in a streambed.

Perceptual Reciprocity in the Neighborhood

It's the reciprocity that matters, enlisting multiple community members to see the world through their senses and experience.

Notes

Chapter 1

1. See the reporting in the *Guardian* from September 27, 2019: https://www.theguardian.com/environment/2019/sep/27/climate-crisis-6-million-people-join-latest-wave-of-worldwide-protests. Moveon.org reported 7.6 million, 6,100 events, and 185 countries. See https://globalclimatestrike.net.

2. Rosenberg et al., "Decline of the North American Avifauna"; Intergovernmental Panel on Climate Change, *Special Report on the Ocean and Cryosphere in a Changing Climate*. Both the August and September 2019 issues of *National Geographic* provide outstanding coverage.

3. During the late 1960s, many independent, "alternative" magazines sprouted. *Ramparts* was a muckraking, progressive journal. *Crawdaddy* was an indie magazine that covered the underground rock scene. Later, the early issues of *Rolling Stone* served the same purpose. The *Coevolution Quarterly* became the extension of the Whole Earth Catalog.

4. For more information about the impact of the *Whole Earth Catalog*, see Maniaque-Benton, *Whole Earth Field Guide*.

5. To peruse back issues of the *Whole Earth Catalog*, see http://www.wholeearth.com/index.php.

6. To get a good understanding of public attitudes regarding climate change and other environmental issues, see the excellent survey work conducted by Anthony Leiserowitz and his team with the Yale Program on Climate Change Communication, https://climatecommunication.yale.edu.

7. As the International Union for Conservation of Nature puts it, "The Millennium Ecosystem Assessment defines Ecosystem Services as 'the benefits people derive from ecosystems.' Besides provisioning services or goods like food, wood, and other raw materials, plans, animals, fungi and micro-organisms provide essential regulating services such as pollination of crops, prevention of soil erosion and water purification, and a vast array if cultural service, like recreation and sense of place." See https://www.iucn.org/commissions/commission-ecosystem-management/our-work/cems-thematic-groups/ecosystem-services.

8. This is the essence of the environmental justice movement and literature. See, in particular, Nixon, *Slow Violence and the Environmentalism of the Poor*.

Chapter 2

1. Hyde, *A Primer for Forgetting*, 29.
2. If you are unfamiliar with the *I Ching*, please refer to the section "I Ching as Improvisational Knowledge System" in chapter 8, "Improvisational Excellence." For an up-to-date history of the book, see Smith, *The I Ching*.
3. Blofeld, *I Ching* , 138.
4. Balkin, *The Laws of Change*, 213.

Chapter 3

1. See https://www.iau-hesd.net/sites/default/files/documents/pnw_changemakersreport_.pdf.
2. For an excellent overview of how rapidly Seattle and Portland are changing, see these fine graphic atlases: Hatfield, Kempson, and Ross, *Seattleness*; Banos and Shobe, *Portlandness*.
3. To learn more about the meaning of the Anthropocene, its origins as a term, and the controversies that surround its usage, see Subcommission on Quaternary Stratigraphy, http://quaternary.stratigraphy.org/working-groups/anthropocene.
4. There are two excellent English-language biographies of Humboldt. See Walls, *The Passage to Cosmos*; Wulf, *The Invention of Nature*. See also Sachs, *The Humboldt Current*.
5. *Global Change and the Earth System* remains an outstanding overview of environmental change science.
6. See the excellent Green Biz network, https://www.greenbiz.com.
7. See the Urban Sustainability Directors Network, https://www.usdn.org/home.html?returnUrl=%2findex.html.
8. All quotations are from page 5 of the report's introduction. See http://www.internal-displacement.org/sites/default/files/publications/documents/2019-IDMC-GRID.pdf.
9. See https://ejfoundation.org/reports/beyond-borders.
10. I highly recommend Dina Gilio-Whitaker's *As Long as Grass Grows*, a powerful assessment of settler colonialism, Native American structural genocide, and the North American history of displacement, deracination, and what she terms "indigenous environmental injustice." She makes the important point that in 1492, there were an estimated 18 million indigenous people in North America, reduced to 228,000 by 1890—a 99 percent decrease. Displacement has a long history in North America, and the consequences are devastating.
11. See Ceballos, Ehrlich, and Dirzo, "Biological Annihilation via the Ongoing Sixth Mass Extinction."
12. Wilson makes this analogy in his classic work *The Diversity of Life*. A more recent work, *Half-Earth*, is an excellent review of the biodiversity crisis. See also Kohlbert, *The Sixth Extinction*.

13. Haddad et al., "Habitat Fragmentation."
14. See https://svs.gsfc.nasa.gov/3827.
15. Sapolsky, *Behave*, 392.
16. I am referring to Greene's book *Moral Tribes*.
17. Sapolsky, *Behave*, 423.
18. Sapolsky, *Behave*, 432.
19. Sapolsky, *Behave*, 291.
20. See https://www.epa.gov/ejscreen.
21. See https://ejatlas.org.
22. Barber, *Cool Cities*, 8.
23. Barber, *Cool Cities*, 10.
24. Snyder, *The Practice of the Wild*, 43.
25. Sapolsky, *Behave*, 424.

Chapter 4

1. The Keeling curve is a graph that measures the accumulation of carbon dioxide in the atmosphere. To learn more about it, see https://scripps.ucsd.edu/programs/keelingcurve.
2. As Alter observes, "We are blind to much that shapes our mental life." Quoted in Brockman, *This Will Make You Smarter*, 150.
3. In a series of books written in the 1960s, McLuhan emphasized that the use of a technology was more important than its content. Hence his famous expression, "The medium is the massage."
4. Crawford, *The World beyond Your Head*. See especially the introduction, "Attention as a Cultural Problem."
5. Kahneman, *Thinking, Fast and Slow*, 20–21.
6. Pagel, *Wired for Culture*, 308, 316.
7. Sapolsky, *Behave*, 674, 673.
8. Meeker, *Minding the Earth*, 3.
9. This famous passage is the last sentence of Darwin's *The Origin of Species*.
10. Trentmann, *Empire of Things*, 1.
11. Deshpande, "Why Curators Matter Now More than Ever."
12. Templeman, "Your Top Seven Tools for Social Media Curation."

Chapter 5

1. White, *Railroaded*, 38.
2. Browne, *Charles Darwin*, 3, 13, 12.
3. The reference "triumph of human empire" refers to the excellent book of that title by Rosalind Williams.

4. In the literature about the Anthropocene, there is discussion about using stratigraphic evidence to locate the distinguishing markers of the Anthropocene, or what some are calling the golden spike. For coverage of this quest, see *Nature*, https://www.nature.com/articles/d41586-019-02381-2.

5. Haskell, *The Songs of Trees*, 38.

6. For a compelling essay, see Vazza and Feletti, "The Strange Similarity of Neuron and Galaxy Networks."

7. Lau, Borrett, Baiser, Gotelli, and Ellison, "Ecological Network Metrics."

8. Landi, Minoarivelo, Brännström, Hui, and Dieckmann, "Complexity and Stability of Ecological Networks.".."

9. Alberti, *Cities That Think like Planets*, 21, 65.

10. Forman, *Land Mosaics*, 5.

11. Lima, *Visual Complexity*, 18.

12. See especially McCandless, *Knowledge Is Beautiful*; Meirelles, *Design for Information*; Borner, *Atlas of Knowledge*; Harmon, *The Map as Art*; Cook, *The Best Information Graphics 2015*; Tufte, *Envisioning Information*.

13. Kadushin, *Understanding Social Networks*, 20.

14. Kadushin, *Understanding Social Networks*, 31.

15. Kadushin, *Understanding Social Networks*, 31, 133.

16. Kadushin, *Understanding Social Networks*, 56.

17. Kadushin, *Understanding Social Networks*, 56, 63.

18. Johnson, *Where Good Ideas Come From*, 51.

Chapter 6

1. Rutherford, *A Brief History of Everyone Who Ever Lived*, 2.

2. For an accessible and delightful discussion, see Hanh, *Interbeing*.

3. All these definitions come from UNHCR, UN Refugee Agency. See https://www.unrefugees.org/refugee-facts/what-is-a-refugee.

4. See https://www.hrw.org/topic/migrants.

5. See http://www.pewresearch.org/fact-tank/2016/12/15/international-migration-key-findings-from-the-u-s-europe-and-the-world.

6. See https://www.un.org/development/desa/publications/international-migration-report-2017.html.

7. See https://www.iom.int.

8. Do cooperation, trade, and exchange lead to tolerance? This is a question worthy of historical and anthropological research. One way to approach it is to read some global histories, although one must be wary of searching for examples that fit predispositions. In reading Joel Kraemer's biography of Moses Maimonides, I was struck by how at various times in twelfth-century Spain and Egypt, economies that thrived on East-West trade were tolerant of diverse religious

perspectives, allowing Muslims, Christians, and Jews surprising autonomy. These eras of tolerance would often give way to cycles of repression and intolerance, spawning conflict, migration and exile. What are the dynamics that cause such shifts?

9. For more on the relationship between ecological and cultural diversity, see UNESCO, http://www.unesco.org/new/en/natural-sciences/environment/ecological-sciences/biodiversity/science-and-research-for-management-and-policy/biological-and-cultural-diversity.

10. Wilcove, *No Way Home*, 2.

11. Vermeij, *Nature*, 52.

12. Vermeij, *The Evolutionary World*, 195.

13. Wilcove, *No Way Home*, 12.

14. Reich, *Who We Are and How We Got Here*, 277.

15. Pagel, *Wired for Culture*, 36.

16. Pagel, *Wired for Culture*, 37.

17. Pagel, *Wired for Culture*, viii.

18. See https://genographic.nationalgeographic.com/human-journey.

19. For a quantitative and visual illustration of these flows, see Abel and Sander, "Quantifying Global International Migration Flows."

20. Saunders, *Arrival City*, 38–39.

21. Saunders, *Arrival City*, 20.

22. Knox, *Atlas of Cities*, 180.

23. See Globalization and World Cities Research Network, https://www.lboro.ac.uk/gawc. Wikipedia re-creates the list with convenient explanations of the city types and colorful flag icons next to the cities. See https://en.wikipedia.org/wiki/Global_city.

24. See https://www.un.org/development/desa/en/news/population/2018-revision-of-world-urbanization-prospects.html.

25. Jordan, "Historical Origins of the One-Drop Racial Rule in the United States," 105.

26. Jordan, "Historical Origins of the One-Drop Racial Rule in the United States," 106.

27. Rutherford, *A Brief History of Everyone Who Ever Lived*, 261.

28. See Yong, *I Contain Multitudes*.

29. UN Refugee Agency, *Left Behind*.

30. Tesh, Burnett, and Nash, *Building Welcoming Schools*.

31. Here's a brief snapshot of children's books addressing migration: Melissa Fleming, *A Hope More Powerful Than the Sea*; Alan Gratz, *Refugee*; Warren St. John, *Outcasts United: The Story of a Refugee Soccer Team That Changed a Town*; Margaret Ruurs and Fallah Raheem, *Stepping Stones: A Refugee Family's*

Journey; Mary Beth Leathersale and Eleanor Shakespeare, *Stormy Seas: Stories of Young Boat Refugees*; Francesca Sanna, *The Journey*.

32. Sanderson, *Mannahatta*.

Chapter 7

1. Arthus-Bertrand, *Earth from Above*; Antoniou, Klanten, and Ehmann, *Mind the Map*; Harmon, *The Map as Art*; Benson, *Cosmigraphics*.
2. For more details on the sense-of-place map, see Thomashow, *Ecological Identity*, 192–199.
3. Dodge, "Living by Life," 6.
4. Thomashow, "Toward a Cosmopolitan Bioregionalism," 121.
5. See https://en.wikipedia.org/wiki/List_of_indigenous_peoples.
6. Dowie, *The Haida Gwaii Lesson*, xiii.
7. Dowie, *The Haida Gwaii Lesson*, xiii.
8. See especially Loh and Harmon, *Biological Diversity*. For an academic overview, see Pretty et al., "The Intersections of Biological and Cultural Diversity.".
9. Dowie, *The Haida Gwaii Lesson*, xiv.
10. See http://www.landmarkmap.org/map/#x=-102.46&y=13.47&l=3&a=community_FormalDoc%2Ccommunity_NoDoc%2Ccommunity_FormalClaim%2Ccommunity_Occupied%2Cindigenous_FormalDoc%2Cindigenous_NoDoc%2Cindigenous_FormalClaim%2Cindigenous_Occupied.
11. Questions of repatriation, sovereignty, and cultural integrity are profoundly important and complex. Dina Gilio-Whitaker's *As Long as Grass Grows: The Indigenous Fight for Environmental Justice, from Colonization to Standing Rock* is a must read for anyone interested in colonization, environmental activism, and indigenous sovereignty.
12. Dowie, *The Haida Gwaii Lesson*, xiv.
13. See https://www.google.com/search?client=safari&rls=en&ei=LZVQW_TcC43j_Abw65dI&q=border+etymology&oq=border&gs_l=psy-ab.1.1.0i67k1l3j0l3j0i67k1j0l2j0i67k1.65887217.0.8562.4.4.0.0.0.0.118.418.2j2.4.0..2..0...1.1.64.psy-ab..0.4.418....0.SuIswNG-vyM.
14. Strayer et al., "A Classification of Ecological Boundaries."
15. Kolasa, "Ecological Boundaries," 35.
16. See https://www.dailymotion.com/video/x1i1e6h.
17. See https://openborders.info.
18. See https://en.wikipedia.org/wiki/List_of_contemporary_ethnic_groups.
19. Dowie, *The Haida Gwaii Lesson*, xv.
20. For the appropriate excerpts from Greene's book, see http://www.thewisepath.org/papers/Six%20Rules%20For%20Modern%20Herders.pdf. Sapolsky, *Behave*, 418–424.
21. See Forbes, *Finding Balance at the Speed of Trust*, 8.

22. Barber, *If Mayors Ruled the World*, 5.
23. Sanderson, *Mannahatta*, 10.
24. Sanderson, *Mannahatta*, 10, 204.
25. Alberti, *Cities That Think like Planets*, 11.
26. Alberti, *Cities That Think like Planets*, 11, 28.
27. Alberti, *Cities That Think like Planets*, 47, 132.
28. Valentine, "Living with Difference," 323.
29. Cohen and Sheringham, *Encountering Difference*, 8–9.
30. Cohen and Sheringham, *Encountering Difference*, 15, 156.
31. Cohen and Sheringham, *Encountering Difference*, 156.
32. Cohen and Sheringham, *Encountering Difference*, 156.
33. Quoted in Cohen and Sheringham, *Encountering Difference*, 40.
34. Kaplan, Sanchez, and Hoffman, *Intergenerational Pathways to a Sustainable Society*, 17.
35. Tsing, *The Mushroom at the End of the World*, 20.
36. Tsing, *The Mushroom at the End of the World*, 27–30.
37. Appiah, *Cosmopolitanism*, 8.

Chapter 8

1. Steffen et al., *Global Change and the Earth System*, 255, 262.
2. For a wonderful cultural and natural history of beavers along with their powerful role in engineering the US landscape, see Goldfarb, *Eager*.
3. She describes the lives of five women as improvisational compositions. Bateson, *Composing a Life*.
4. Ayto, *Dictionary of Word Origins*, 296.
5. Macfarlane, *Landmarks*, 3.
6. Macfarlane, *Landmarks*, 1.
7. Goodnough, "Survivors of Tsunami Live on Close Terms with Sea."
8. Macfarlane, *Landmarks*, 13.
9. Macfarlane, *Landmarks*, 14.
10. Lukas, *Language Making Nature*, 77.
11. Nikula et al., *Stokes Beginner's Guide to Dragonflies*, 149.
12. Macfarlane, *Landmarks*, 346.
13. These discussions were published in a commemorative brochure celebrating Tom's work at Antioch University New England. It is no longer available. Please contact me if you would like to see the full interview.
14. Pretor-Pinney, *The Wave Watcher's Companion*, 17.
15. Golley, *How to Read Water*, 1.
16. Allen and Hoekstra, *Toward a Unified Ecology*, 398.

17. See also Ball, *The Self-Made Tapestry*.

18. Of special interest is volume 1 of *The Nature of Order, The Phenomenon of Life*, chapter 6, "The Fifteen Properties in Nature," 244–296.

19. I first read *Bartholomew and the Oobleck* as a young child and it made a lasting impression. The king was bored with the weather so he asked his wizards to create an unprecedented form of precipitation. They created a green sticky ooze that stuck to everything and had no idea how to stop it. How pertinent!

20. These definitions are all derived from Steffen et al., *Global Change and the Earth System*, 265–267.

21. See https://plato.stanford.edu/entries/chance-randomness.

22. See https://www.urbandictionary.com/define.php?term=Random.

23. For comments on uncertainty, emergent property, and random processes in game design, see Salen and Zimmerman, *Rules of Play*.

24. Impey, *Humble before the Void*, 146.

25. Burghardt, *The Genesis of Animal Play*, xii.

26. Nachmanovitch, *Free Play*, 33.

27. Vermeij, *The Evolutionary World*, 20.

28. Some places to start include Asma, *The Evolution of Imagination*; Johnson, *Where Good Ideas Come From*; Prum, *The Evolution of Beauty*; Rothenberg, *Survival of the Beautiful*.

29. Prum, *The Evolution of Beauty*, 336.

30. See all three books by Rothenberg: *Why Birds Sing*, *Thousand Mile Song*, and *Bug Music*.

31. Asma, *The Evolution of Imagination*, 197.

32. Polymath, https://polymathprojects.org; GenBank, https://www.ncbi.nlm.nih.gov/genbank; Galaxy Zoo, https://blog.galaxyzoo.org.

33. Nielsen, *Reinventing Discovery*, 30.

34. NEON, https://www.neonscience.org/observatory/about.

35. For a history of the establishment of LTER's research stations, see https://lternet.edu/network-organization/lter-a-history.

36. From the 2018 LTER conference agenda. See https://lternet.edu/lter-scientists-meeting-2018.

37. Asma, *The Evolution of Imagination*, 89.

38. Ming-Dao, *The Living I Ching*, 33.

39. Ming-Dao, *The Living I Ching*, 33.

40. Minford, *I Ching*, xxiv.

41. Minford, *I Ching*, xxvi. Minford says more about Fu Xi: "Traditionally the invention of these Hexagrams, or rather of the three-line Trigrams that were thought to constitute them, was ascribed to the legendary Fux Xi, divinely inspired by his observations of the Patterns of the Universe, of Nature, of Heaven

and Earth." Minford, *I Ching*, 10. Evidence suggests that Fu Xi lived around 2850 BC. Hence sixteen hundred years of ancient Chinese history separates Fu Xi from King Wen.

42. Minford, *I Ching*, 130–136.
43. Balkin, *The Laws of Change*, 216–222.
44. See https://www.forestkindergartenassociation.org.
45. Nachmanovitch, *Free Play*, 27.

Chapter 9

1. See https://www.etymonline.com/word/glimpse.
2. Abraham Joshua Heschel writes at great length about the ineffable in *Man Is Not Alone*.
3. Pollan, *How to Change Your Mind*, 16.
4. Pollan, *How to Change Your Mind*, 15.
5. Pollan, *How to Change Your Mind*, 16.
6. For more on Romantic science, see Holmes, *The Age of Wonder*. For a biography of Humboldt, see Wulf, *The Invention of Nature*.
7. Pollan, *How to Change Your Mind*, 128.
8. Pollan, *How to Change Your Mind*, 318–319.
9. Pollan, *How to Change Your Mind*, 308–309.
10. Von Uexküll, *A Foray into the Worlds of Animals and Humans*, 43.
11. Cited in von Uexküll, *A Foray into the Worlds of Animals and Humans*, 9, 21.
12. See https://www.garrisoninstitute.org/blog/insight-of-interbeing.
13. Deming, *Zoologies*, 7.
14. Powers, *The Overstory*, 283.
15. Haskell, *The Songs of Trees*, 9.
16. For a wonderful series of essays on relationships with trees, see Kaza, *The Attentive Heart*.
17. Powers, *The Overstory*, 282.
18. Yong, *I Contain Multitudes*, 2–3, 17.
19. Yong, *I Contain Multitudes*, 19, 26.
20. Margulis and Sagan, *What Is Life?*, 176.
21. Sanderson, *Mannahatta*, 33.
22. Haskell, *The Songs of Trees*, 37, 35.
23. Horowitz, *On Looking*, 2–3, 53.
24. See https://www.pps.org.
25. See https://www.barefootcollege.org.

Bibliography

Abel, Guy J., and Nikola Sander. "Quantifying Global International Migration Flows." *Science*, March 28, 2014.

Aberley, Doug, ed. *Boundaries of Home: Mapping for Local Empowerment*. New Society, 1993.

Abram, David. *Becoming Animal: An Earthly Cosmology*. Pantheon Books, 2010.

Abram, David. "Creaturely Migrations on a Breathing Planet." *Emergence Magazine* 1. https://emergencemagazine.org/story/creaturely-migrations-breathing-planet.

Agyeman, Julian. *Introducing Just Sustainabilities: Policy, Planning, and Practice*. Zed Books, 2016.

Alberti, Marina. *Cities That Think like Planets: Complexity, Resilience, and Innovation in Hybrid Ecosystems*. University of Washington Press, 2016.

Alexander, Christopher. *A Pattern Language*. Oxford University Press, 1977.

Alexander, Christopher. *The Nature of Order: An Essay on the Art of Building and the Nature of the Universe. Book One: The Phenomenon of Life*. Center for Environmental Structure, 2002.

Alexander, Stephon. *The Jazz of Physics: The Secret Link between Music and the Structure of the Universe*. Basic Books, 2016.

Allen, Timothy F. H., and Thomas W. Hoekstra. *Toward a Unified Ecology*. Columbia University Press, 2015.

Alter, Adam. *Irresistible: The Rise of Addictive Technology and the Business of Keeping Us Hooked*. Penguin Press, 2017.

Antoniou, Antonis, Robert Klanten, and Sven Ehmann, eds. *Mind the Map: Illustrated Maps and Cartography*. Gestalten, 2015.

Appiah, Kwame Anthony. *Cosmopolitanism: Ethics in a World of Strangers*. W. W. Norton and Company, 2006.

Arthus-Bertrand, Yann. *Earth from Above*. Abrams, 1999.

Asma, Stephen T. *The Evolution of Imagination*. University of Chicago Press, 2017.

Ayto, John. *Dictionary of Word Origins*. Arcade Publishing, 1990.

Balkin, Jack. *The Laws of Change: I Ching and the Philosophy of Life*. Schocken Books, 2002.

Ball, Philip. *Patterns in Nature: Why the Natural World Looks the Way It Does*. University of Chicago Press, 2016.

Ball, Philip. *The Self-Made Tapestry: Pattern Formation in Nature*. Oxford University Press, 1999.

Banos, David, and Hunter Shobe. *Portlandness: A Cultural Atlas*. Sasquatch Books, 2015.

Barber, Benjamin R. *Cool Cities: Urban Sovereignty and the Fix for Global Warming*. Yale University Press, 2017.

Barber, Benjamin R. *If Mayors Ruled the World: Dysfunctional Nations, Rising Cities*. Yale University Press, 2013.

Barber, Benjamin R. *Strong Democracy: Participatory Politics for a New Age*. University of California Press, 1984.

Bateson, Mary Catherine. *Composing a Life*. Grove Press, 2001.

Behrmann, Simon, and Avidan Kent, eds. *Climate Refugees*. Routledge, 2018.

Benson, Michael. *Cosmigraphics: Picturing Space through Time*. Abrams, 2014.

Berliner, Paul F. *Thinking in Jazz: The Infinite Art of Improvisation*. University of Chicago Press, 1994.

Blofeld, John. *I Ching: The Book of Change*. E. P. Dutton, 1968.

Bonnett, Alastair. *An Uncommon Atlas: 50 New Views of Our Physical, Cultural, and Political World*. White Lion Publishing, 2019.

Borner, Katy. *Atlas of Knowledge: Anyone Can Map*. MIT Press, 2015.

Brockman, John, ed. *This Will Make You Smarter: New Concepts to Improve Your Thinking*. Harper Perennial, 2012.

Browne, Janet. *Charles Darwin: The Power of Place*. Alfred A. Knopf, 2002.

Bullard, Robert D. *Unequal Protection: Environmental Justice and Communities of Color*. Random House, 1994.

Burghardt, Gordon M. *The Genesis of Animal Play: Testing the Limits*. MIT Press, 2005.

Carhart-Harris, Robin L., Robert Leech, Peter J. Hellyer, Murray Shanahan, Amanda Feilding, Enzo Tagliazucchi, Dante R. Chiavo, and David Nutt. "The Entropic Brain: A Theory of Conscious States Informed by Neuroimaging Research with Psychedelic Drugs." *Frontiers in Human Neuroscience*, February 3, 2014.

Carr, Nicholas G. "What Scientific Concept Would Improve Everybody's Cognitive Toolkit?" *Edge*, 2011. https://www.edge.org/response-detail/10220.

Ceballos, Gerardo, Paul R. Ehrlich, and Rodolfo Dirzo. "Biological Annihilation via the Ongoing Sixth Mass Extinction Signaled by Vertebrate Population Losses and Declines." *Proceedings of the National Academy of Sciences*, July 10, 2017. https://doi.org/10.1073/pnas.1704949114.

Cheshire, James, and Oliver Uberti. *Where the Animals Go: Tracking Wildlife with Technology in 50 Maps and Graphics*. W. W. Norton and Company, 2016.

Chirico, Jennifer, and Gregory S. Farley. *Thinking like an Island: Navigating a Sustainable Future in Hawai'i*. University of Hawai'i Press, 2015.

Cleary, Thomas. *I Ching Mandalas: A Program of Study for the Book of Changes*. Shambhala, 1989.

Cohen, Robin, and Olivia Sheringham. *Encountering Difference: Diasporic Traces, Creolizing Spaces*. Polity Press, 2016.

Collectif Argos. *Climate Refugees*. MIT Press, 2010.

Cook, Gareth, ed. *The Best Information Graphics 2015*. Houghton Mifflin Harcourt, 2015.

Crawford, Matthew B. *The World beyond Your Head: On Becoming an Individual in an Age of Distraction*. Farrar, Straus and Giroux, 2015.

Csikszentmihalyi, Mihaly. *Flow: The Psychology of Optimal Experience*. Harper Perennial Modern Classics, 2008.

Darwin, Charles. *The Origin of Species*. Penguin Classics, 1985.

Deming, Alison H. *Zoologies: On Animals and the Human Spirit*. Milkweed, 2014.

Deming, Alison H., and Lauret E. Savoy. *Colors of Nature: Culture, Identity and the Natural World*. Milkweed, 2011.

Deshpande, Pawan. "Why Curators Matter Now More than Ever." *Huffington Post*, October 3, 2013. https://www.huffpost.com/entry/why-curators-matter-now-m_b_4037594.

Dodge, Jim. "Living by Life: Some Bioregional Theory and Practice." *Coevolution Quarterly* 32 (Winter 1981): 6–12.

Dodge, Jim. *Rain on the River*. Grove Press, 2002.

Dowie, Mark. *The Haida Gwaii Lesson: A Strategic Playbook for Indigenous Sovereignty*. Inkshares, 2017.

Erdrich, Heid E., ed. *New Poets of Native Nations*. Greywolf Press, 2018.

Ferguson, Niall. *The Square and the Tower: Networks and Power, from the Freemasons to Facebook*. Penguin Press, 2018.

Flannery, Kent, and Joyce Marcus. *The Creation of Inequality: How Our Prehistoric Ancestors Set the Stage for Monarchy, Slavery, and Empire*. Harvard University Press, 2012.

Forbes, Peter. *Finding Balance at the Speed of Trust: The Story of Southeast Sustainable Partnerships*. https://www.nature.org/content/dam/tnc/nature/en/documents/SSP-Speed-of-Trust.pdf.

Forman, Richard T. T. *Land Mosaics: The Ecology of Landscapes and Regions*. Cambridge University Press, 1995.

Frankopan, Peter. *The Silk Roads: A New History of the World*. Alfred A. Knopf, 2016.

Geisel, Theodor (Dr. Seuss). *Bartholomew and the Oobleck*. Random House, 1949.

Ghosh, Amitav. *The Great Derangement: Climate Change and the Unthinkable*. University of Chicago Press, 2016.

Ghosh, Amitav. *Gun Island*. Farrar, Straus and Giroux, 2019.

Gilio-Whitaker, Dina. *As Long as Grass Grows: The Indigenous Fight for Environmental Justice, from Colonization to Standing Rock*. Beacon Press, 2019.

Godfrey-Smith, Peter. *Other Minds: The Octopus and the Evolution of Intelligent Life*. Farrar, Straus and Giroux, 2017.

Goldfarb, Ben. *Eager: The Surprising Story of Beavers and Why They Matter*. Chelsea Green, 2019.

Golley, Tristan. *How to Read Water: Clues and Patterns from Puddles to the Sea*. Workman Publishing, 2016.

Goodnough, Abby. "Survivors of Tsunami Live on Close Terms with Sea." *New York Times*, January 23, 2005.

Govinda, Lama Anagarika. *The Inner Structure of the I Ching: The Book of Transformations*. Wheelwright Press, 1981.

Greene, Joshua. *Moral Tribes: Emotion, Reason, and the Gap between Us and Them*. Penguin Books, 2014.

Haddad, Nick M., Lars A. Brudvig, Jean Clobert, Kendi F. Davies, Andrew Gonzalez, Robert D. Holt, Thomas E. Lovejoy, et al. "Habitat Fragmentation and Its Lasting Impact on Earth Systems." *Science Advances* (2015): 1:e1500052. https://www.researchgate.net/publication/274389394_Habitat_fragmentation_and_its_lasting_impact_on_Earth_ecosystems.

Hall, Stephen S. *Mapping the Next Millennium*. Random House, 1992.

Hall, Stephen S. *Wisdom: From Philosophy to Neuroscience*. Alfred A. Knopf, 2010.

Hamid, Moshin. *Exit West*. Random House, 2017.

Hanh, Thich Nhat. *Interbeing: Fourteen Guidelines for Engaged Buddhism*. Parallax Press, 1987.

Harari, Yuval Noah. *Homo Deus: A Brief History of Tomorrow*. Harper Perennial, 2018.

Harmon, Katharine. *The Map as Art: Contemporary Artists Explore Cartography*. Princeton Architectural Press, 2009.

Harmon, Katharine. *You Are Here: NYC: Mapping the Soul of the City*. Princeton Architectural Press, 2017.

Haskell, David George. *The Songs of Trees: Stories from Nature's Great Connectors*. Penguin Random House, 2017.

Hatfield, Tera, Jenny Kempson, and Nathalie Ross. *Seattleness: A Cultural Atlas*. Sasquatch Books, 2018.

Hay, Pete. *Physick*. Shoestring Press, 2016.

Heise, Ursula K. *Sense of Place and Sense of Planet: The Environmental Imagination of the Global.* Oxford University Press, 2008.

Henderson, Caspar. *The Book of Barely Imagined Beings: A 21st Century Bestiary.* University of Chicago Press, 2013.

Henderson, Caspar. *A New Map of Wonders: A Journey in Search of Modern Marvels.* University of Chicago Press, 2017.

Hersey, Mark D. *My Work Is That of Conservation: An Environmental Biography of George Washington Carver.* University of Georgia, 2011.

Heschel, Abraham Joshua. *Man Is Not Alone: A Philosophy of Religion.* Farrar, Straus and Giroux, 1951.

Hinton, David. *I Ching: The Book of Change.* Farrar, Straus and Giroux, 2015.

Holmes, Richard. *The Age of Wonder: How the Romantic Generation Discovered the Beauty and Terror of Science.* Random House, 2008.

Horowitz, Alexandra. *On Looking: Eleven Walk with Expert Eyes.* Scribner, 2013.

Hughes, Howard C. *Sensory Exotica: A World beyond Human Experience.* MIT Press, 1999.

Hyde, Lewis, ed. *The Essays of Henry D. Thoreau.* Farrar, Straus and Giroux, 2002.

Hyde, Lewis. *The Gift: How the Creative Spirit Transforms the World.* Random House, 1983.

Hyde, Lewis. *A Primer for Forgetting: Getting Past the Past.* Farrar, Straus and Giroux, 2019.

Impey, Chris. *Humble before the Void: A Western Astronomer, His Journey East, and a Remarkable Encounter between Western Science and Tibetan Buddhism.* Templeton Press, 2014.

Intergovernmental Panel on Climate Change. *Special Report on the Ocean and Cryosphere in a Changing Climate.* https://www.ipcc.ch/srocc/download-report.

Jablonski, Eva, and Marion J. Lamb. *Evolution in Four Dimensions: Genetic, Epigenetic, Behavioral, and Symbolic Variation in the History of Life.* MIT Press, 2014.

Jahren, Hope. *Lab Girl.* Penguin Random House, 2017.

Johnson, Crockett. *Harold and the Purple Crayon.* Harper and Brothers, 1955.

Johnson, Steven. *Where Good Ideas Come From: The Natural History of Innovation.* Penguin Group, 2010.

Jordan, Winthrop D. "Historical Origins of the One-Drop Racial Rule in the United States." *Journal of Critical Mixed Race Studies* 1, no. 1 (2014): 98–132.

Kadushin, Charles. *Understanding Social Networks: Theories, Concepts, and Findings.* Oxford University Press, 2012.

Kahneman, Daniel. *Thinking, Fast and Slow.* Farrar, Straus and Giroux, 2011.

Kaplan, Matthew, Mariano Sanchez, and Jacobus Hoffman. *Intergenerational Pathways to a Sustainable Society.* Springer, 2017.

Kaufman, Kenn. *A Season on the Wind: Inside the World of Spring Migration.* Boston: Houghton Mifflin Harcourt, 2019.

Kaza, Stephanie. *The Attentive Heart: Conversations with Trees.* Fawcett, 1993.

Khanna, Parag. *Connectography: Mapping the Future of Global Civilization.* Random House, 2016.

Kimmerer, Robin Wall. *Braiding Sweetgrass: Indigenous Wisdom, Scientific Knowledge, and the Teachings of Plants.* Milkweed, 2013.

Klein, Naomi. *On Fire: The Burning Case for a Green New Deal.* Simon and Schuster, 2019.

Klein, Naomi. *This Changes Everything: Capitalism vs. the Climate.* Simon and Schuster, 2015.

Knox, Paul, ed. *Atlas of Cities.* Princeton University Press, 2014.

Kohlbert, Elizabeth. *The Sixth Extinction: An Unnatural History.* Picador, 2015.

Kohn, Eduardo. *How Forests Think: Toward an Anthropology beyond the Human.* University of California Press, 2013.

Kolasa, J. "Ecological Boundaries: A Derivative of Ecological Entities." *Web Ecology* 14, no. 1 (2014): 27–37.

Kraemer, Joel. L. *Maimonides: The Life and Times of One of the World's Greatest Minds.* Doubleday, 2008.

Krause, Bernie. *The Great Animal Orchestra: Finding the Origins of Music in the World's Wild Places.* Little, Brown and Company, 2012.

Krause, Bernie. *Wild Soundscapes: Discovering the Voice of the Natural World.* Wilderness Press, 2002.

Laland, Kevin N. *Darwin's Unfinished Symphony: How Culture Made the Human Mind.* Princeton University Press, 2017.

Landi, Pietro, Henintsoa O. Minoarivelao, Åke Brännström, Canh Hui, and Ulf Dieckmann. "Complexity and Stability of Ecological Networks: A Review of the Theory." *Population Ecology* (2018): 60.

Lau, M. K., S. R. Borrett, B. Baiser, N. J. Gotelli, and A. M. Ellison. "Ecological Network Metrics: Opportunities for Synthesis." *Ecosphere* 8 (2017): e)1900. 10.1002/ecs2.1900.

Lear, Linda. *Rachel Carson: Witness for Nature.* Henry Holt and Company, 1997.

Lenton, Tim, and Andrew Watson. *Revolutions That Made the Earth.* Oxford University Press, 2011.

Lima, Manuel. *Visual Complexity: Mapping Patterns of Information.* Princeton Architectural Press, 2011.

Loh, Jonathan, and David Harmon. *Biological Diversity: Threatened Species, Endangered Languages.* WWF Netherlands, 2014. http://wwf.panda.org/wwf_news/222890/Biocultural-Diversity-Threatened-Species-Endangered-Languages.

Lukas, David. *Language Making Nature: A Handbook for Artists, Writers and Thinkers.* Lukas Guides, 2015.

Lynch, Tom, Cheryl Glotfelty, and Karla Armbruster. *The Bioregional Imagination*. University of Georgia Press, 2012.

Macfarlane, Robert. *Landmarks*. Penguin Books, 2015.

Maier, Charles S. *Once within Borders: Territories of Power, Wealth, and Belonging since 1500*. Harvard University Press, 2016.

Maniaque-Benton, Caroline, ed. *Whole Earth Field Guide*. MIT Press, 2016.

Mann, Thomas. *Joseph and His Brothers*. Everyman's Library, 2005.

Margulis, Lynn, and Dorion Sagan. *Microcosmos: Four Billion Years of Microbial Evolution*. University of California Press, 1986.

Margulis, Lynn, and Dorion Sagan. *What Is Life?* Simon and Schuster, 1995.

Margulis, Lynn, Karlene V. Schwartz, and Michael Dolan. *Diversity of Life: An Illustrated Guide to the Five Kingdoms*. Jones and Bartlett, 1999.

Mathieu, W. A. *Bridge of Waves: What Music Is and How Listening to It Changes the World*. Shambhala, 2010.

McCandless, David. *Knowledge Is Beautiful*. HarperCollins, 2014.

McGinnis, Michael Vincent, ed. *Bioregionalism*. Routledge, 1999.

McKibben, Bill. *Falter: Has the Human Game Begun to Play Itself Out?* Henry Holt and Company, 2019.

Meeker, Joseph W. *Minding the Earth: Thinly Disguised Essays on Human Ecology*. Latham Foundation, 1988.

Meirelles, Isabel. *Design for Information: An Introduction to the Histories, Theories, and Best Practices behind Effective Information Visualization*. Rockport Publishers, 2013.

Miller, Todd. *Storming the Wall: Climate Change, Migration, and Homeland Security*. City Lights Books, 2017.

Minford, John. *I Ching*. Penguin Books, 2014.

Ming-Dao, Deng. *The Living I Ching: Using Ancient Chinese Wisdom to Shape Your Life*. HarperCollins, 2006.

Montgomery, Sy. *How to Be a Good Creature: A Memoir in Thirteen Animals*. Houghton Mifflin Harcourt, 2018.

Moore, Kathleen Dean. "A Field Guide to Western Birds." In *Companions on Wonder: Children and Adults Exploring Nature Together*, edited by Julie Dunlap and Stephen Kellert. MIT Press, 2012.

Mostafavi, Mohsen, and Gareth Doherty, eds. *Ecological Urbanism*. Lars Muller Publishers, 2016.

Nachmanovitch, Stephen. *Free Play: Improvisation in Life and Art*. Penguin Putnam, 1990.

Nelson, Richard. *The Island Within*. North Point Press, 1989.

Nielsen, Michael. *Reinventing Discovery: The New Era of Networked Science*. Princeton University Press, 2011.

Nikula, Blair, Jackie Sones, Donald Stokes, and Lillian Stokes. *Stokes Beginner's Guide to Dragonflies*. Little, Brown and Company, 2002.

Nixon, Rob, *Slow Violence and the Environmentalism of the Poor*. Harvard University Press, 2013.

Pagel, Mark. *Wired for Culture: Origins of the Human Social Mind*. W. W. Norton and Company, 2012.

Pemberton, Simon, and Jenny Phillimore. "Migrant Place-Making in Super-Diverse Neighbourhoods: Moving beyond Ethno-National Approaches." *Urban Studies*, July 5, 2016. http://pure-oai.bham.ac.uk/ws/files/29213189/PDF_Proofu Ustudies.PDF.

Petri, Giovanni, Paul Expert, Frederico Turkheimer, Robin L. Carhart-Harris, David Nutt, Peter J. Hellyer, and Franco Vaccarino. "Homological Scaffolds of Brain Functional Networks." *Journal of the Royal Society Interface*, December 6, 2014.

Piketty, Thomas. *Capital in the Twenty-First Century*. Harvard University Press, 2017.

Pollan, Michael. *How to Change Your Mind: What the New Science of Psychedelics Teaches Us about Consciousness, Dying, Addiction, Depression, and Transcendence*. Penguin Press, 2018.

Powers, Richard. *The Overstory*. W. W. Norton and Company, 2018.

Pretor-Pinney, Gavin. *The Wave Watcher's Companion: From Ocean Waves to Light Waves via Shock Waves, Stadium Waves, and All the Rest of Life's Undulations*. Penguin Group, 2010.

Pretty, Jules, Bill Adams, Fikret Berkes, Simone Ferreira de Athayde, Nigel Dudley, Eugene Hunn, Luisa Maffi, et. al. "The Intersections of Biological and Cultural Diversity: Towards Integration." *Conservation and Society* 7, no. 2 (2009): 100–112.

Prum, Richard O. *The Evolution of Beauty: How Darwin's Forgotten Theory of Mate Choice Shapes the Animal World—and Us*. Penguin Random House, 2017.

Rajan, S. Irudaya, and R. B. Bhagat, eds. *Climate Change, Vulnerability, and Migration*. Routledge India, 2019.

Redfern, Ron. *Origins: The Evolution of Continents, Oceans, and Life*. University of Oklahoma Press, 2001.

Reich, David. *Who We Are and How We Got Here: Ancient DNA and the New Science of the Human Past*. Pantheon Books, 2018.

Rendgen, Sarah, and Julius Wiedermann. *Understanding the World: The Atlas of Infographics*. Taschen, 2018.

Ritsema, Rudolf, and Shantena Augusto Sabbadini. *The Original I Ching Oracle*. Watkins Publishing, 2005.

Robinson, Mary. *Climate Justice: Hope, Resistance, and the Fight for a Sustainable Future*. Bloomsbury Publishing, 2018.

Rogers, Pattiann. *Generations*. Penguin Group, 2004.

Ronnberg, Amy, and Kathleen Morris, eds. *The Book of Symbols: Reflections on Archetypal Images*. Taschen, 2010.

Rosenberg, Kenneth V., Adriaan M. Dokter, Peter J. Blancher, John R. Sauer, Adam C. Smith, Paul A. Smith, Jessica C. Stanton, Arvind Panjabi, Laura Helft, Michael Parr, and Peter P. Marra. "Decline of the North American Avifauna." *Science* 366, no. 6461 (October 4, 2019): 120–124.

Rothenberg, David. *Bug Music: How Insects Gave Us Rhythm and Noise*. St. Martin's Press, 2013.

Rothenberg, David. *Sudden Music*. University of Georgia Press, 2002.

Rothenberg, David. *Survival of the Beautiful: Art, Science, and Evolution*. Bloomsbury Press, 2011.

Rothenberg, David. *Thousand Mile Song: Whale Music in a Sea of Sound*. Basic Books, 2008.

Rothenberg, David. *Why Birds Sing: A Journey into the Mystery of Bird Song*. Basic Books, 2005.

Rovelli, Carlo. *The Order of Time*. Riverhead Books, 2018.

Rutherford, Adam. *A Brief History of Everyone Who Ever Lived: The Human Story Retold through Genes*. Experiment, 2017.

Sachs, Aaron. *The Humboldt Current: Nineteenth-Century Exploration and the Roots of American Environmentalism*. Viking, 2006.

Safina, Carl. *Beyond Words: What Animals Think and Feel*. Henry Holt and Company, 2015.

Sagan, Dorion, and Lynn Margulis. *Garden of Microbial Delights: A Practical Guide to the Subvisible World*. Kendall Hunt, 1993.

Salen, Katie, and Eric Zimmerman. *Rules of Play: Game Design Fundamentals*. MIT Press, 2003.

Sanderson, Eric W. *Mannahatta: A Natural History of New York City*. Abrams, 2009.

Sapolsky, Robert. *Behave: The Biology of Humans at Our Best and Worst*. Penguin Press, 2017.

Saunders, Doug. *Arrival City: How the Largest Migration in History Is Reshaping Our World*. Random House, 2010.

Savoy, Lauret E. *Trace: Memory, History, Race and the American Landscape*. Counterpoint, 2015.

Schwenk, Theodor. *Sensitive Chaos: The Creation of Flowing Forms in Water and Air*. Rudolf Steiner Press, 1965.

Shi, David E. *The Simple Life: Plain Living and High Thinking in American Culture*. Oxford University Press, 1985.

Smil, Vaclav. *The Earth's Biosphere: Evolution, Dynamics, and Change*. MIT Press, 2002.

Smith, Annick, and Susan O'Connor. *Hearth: A Global Conversation on Community, Identity, and Place.* Milkweed, 2018.

Smith, Anthony D. *The Ethnic Origins of Nations.* Basil Blackwell, 1986.

Smith, Richard J. *The I Ching: A Biography.* Princeton University Press, 2012.

Snyder, Gary. *The Gary Snyder Reader.* Counterpoint, 1999.

Snyder, Gary. *No Nature.* Pantheon Books, 1992.

Snyder, Gary. *The Practice of the Wild.* North Point Press, 1990.

Snyder, Gary. *This Present Moment.* Counterpoint, 2015.

Snyder, Gary. *Turtle Island.* New Directions, 1969.

Solnit, Rebecca, and Joshua Jelly-Schapiro. *Nonstop Metropolis: A New York City Atlas.* University of California Press, 2016.

Souder, William. *On a Farther Shore: The Life and Legacy of Rachel Carson.* Random House, 2012.

Soseki, Muso. *Sun at Midnight.* North Point Press, 1989.

Stamets, Paul. *Mycelium Running: How Mushrooms Can Help Save the World.* Ten Speed Press, 2005.

Steffen, W., R. A. Sanderson, P. D. Tyson, J. Jäger, P. A. Matson, B. Moore III, F. Oldfield, K. Richardson, H.-J. Schellnhuber, B. L. Turner, and R. J. Watson. *Global Change and the Earth System: A Planet under Pressure.* Springer-Verlag, 2003.

Strayer, David L., Mary E. Power, William F. Afghan, Steward T. A. Pickett, and Jayne Belnap. "A Classification of Ecological Boundaries." *BioScience* 53, no. 8 (August 2003): 723–729.

Taylor, Dorceta E. *The Environment and the People in American Cities, 1600s–1990s: Disorder, Inequality and Social Change.* Duke University Press, 2009.

Tegmark, Mark. *Life 3.0: Being Human in the Age of Artificial Intelligence.* Alfred A. Knopf, 2017.

Templeman, Mike. "Your Top Seven Tools for Social Media Curation." *Forbes*, January 31, 2017. https://www.forbes.com/sites/miketempleman/2017/01/31/your-top-7-tools-for-social-media-content-curation/#6da112951f84.

Tesh, Claire, Sara Burnett, and Andy Nash. *Building Welcoming Schools: A Guide for K-12 Educators and After-School Providers.* Welcoming America, 2017. https://19lwtt3nwtm12axw5e31ay5s-wpengine.netdna-ssl.com/wp-content/uploads/2018/02/WR_K12Toolkit_Final.pdf.

Tharoor, Kanishk. *Swimmer among the Stars.* Farrar, Straus and Giroux, 2017.

Thomashow, Mitchell. *Bringing the Biosphere Home: Learning to Perceive Global Environmental Change.* MIT Press, 2001.

Thomashow, Mitchell. *Ecological Identity: Becoming a Reflective Environmentalist.* MIT Press, 1995.

Thomashow, Mitchell. "Toward a Cosmopolitan Bioregionalism." In *Bioregionalism*, edited by Michael Vincent McGinnis. Routledge, 1999.

Thompson, Evan. *Waking, Dreaming, Being.* Columbia University Press, 2015.

Thorpe, Helen. *The Newcomers: Finding Refuge, Friendship, and Hope in an American Classroom.* Simon and Schuster, 2017.

Trentmann, Frank. *Empire of Things: How We Became a World of Consumers, from the Fifteenth Century to the Twenty-First.* HarperCollins, 2016.

Tsing, Anna Lowenhaupt. *The Mushroom at the End of the World: On the Possibility of Life in Capitalist Ruins.* Princeton University Press, 2015.

Tufte, Edward R. *Envisioning Information.* Graphics Press, 1990.

UN Refugee Agency. *Left Behind: Refugee Education in Crisis.* 2017. https://www.unhcr.org/59b696f44.pdf.

Valentine, Gill. "Living with Difference: Reflections on Geographies of Encounter." *Progress in Human Geography* 32, no. 3 (2008): 323–337.

Vazza, Franco, and Alberto Feletti. "The Strange Similarity of Neuron and Galaxy Networks." *Nautilus*, July 25, 2019. http://nautil.us/issue/50/emergence/the-strange-similarity-of-neuron-and-galaxy-networks.

Vermeij, Geerat J. *The Evolutionary World: How Adaptation Explains Everything from Seashells to Civilization.* St. Martin's Press, 2010.

Vermeij, Geerat J. *Nature: An Economic History.* Princeton University Press, 2004.

Vernardsky, Vladimir. *The Biosphere.* Springer-Verlag, 1998.

Volk, Tyler. *CO_2 Rising: The World's Greatest Environmental Challenge.* MIT Press, 2008.

Volk, Tyler. *Gaia's Body: Toward a Physiology of Earth.* Springer-Verlag, 1998.

Von Uexküll, Jakob. *A Foray into the World of Animals and Humans.* University of Minnesota Press, 2010.

Walls, Laura Dassow. *Henry David Thoreau: A Life.* University of Chicago Press, 2017.

Walls, Laura Dassow. *The Passage to Cosmos.* University of Chicago Press, 2009.

Weber, Andreas. *The Biology of Wonder: Aliveness, Feeling, and the Metamorphosis of Science.* New Society Publishers, 2016.

Wessels, Tom. *Reading the Forested Landscape: A Natural History of New England.* Countryman Press, 1997.

White, Richard. *Railroaded: The Transcontinentals and the Making of Modern America.* W. W. Norton and Company, 2011.

Wilcove, David S. *No Way Home: The Decline of the World's Great Animal Migrations.* Island Press, 2008.

Wilhelm, Richard, and Cary F. Baynes. *I Ching or Book of Changes.* Princeton University Press, 1987.

Williams, Rosalind. *The Triumph of Human Empire: Verne, Morris, and Stevenson at the End of the World.* University of Chicago Press, 2013.

Wilson, E. O. *The Diversity of Life.* Harvard University Press, 1992.

Wilson, E. O. *Half-Earth: Our Planet's Fight for Life*. W. W. Norton and Company, 2016.

Wulf, Alexandra. *The Invention of Nature: Alexander von Humboldt's New World*. Alfred A. Knopf, 2015.

Yong, Ed. *I Contain Multitudes: The Microbes within Us and a Grander View of Life*. HarperCollins, 2016.

Yoon, Carol Kaesuk. *Naming Nature: The Clash between Instinct and Science*. W. W. Norton and Company, 2009.

Zim, Herbert. *The Golden Guide to the Stars*. Simon and Schuster, 1951.

Index

Aberley, Doug, 139, 150
Abram, David, 190
Adaptation, 186–189
Adaptive learning, 27–28
  improvisation and, 186–189
Adaptive management, 186–187
Alberti, Marina, 99–100, 107, 153
Alexander, Christopher, 182
Alexander, Stephon, 190
Allen, Timothy, and Thomas W.
  Hoekstra, 78, 181–182
Alter, Adam, 64
Ancestry, 115–117
Anthropocene
  autonomic response, 62
  global environmental change and,
    160
  first nations and, 47
  history and background, 39–41,
    239
  language of, 178
  psychological ramifications of, 61,
    63–66
Asma, Stephen T., 186, 190, 193,
  195
Attention competition, 66–69
Autonomy, 65, 70, 239
Ayto, John, 173–174

Balkin, Jack, 198
Ball, Philip, 182
Barber, Benjamin, 50–52, 151
Berliner, Paul, 190
Biogeochemical cycles, 45–46, 181

Bioregionalism
  Coevolution Quarterly, 138–139
  concept of, 150–152
  cosmopolitan, 139, 150
  indigenous cultures and, 140–142,
    151
  political strategy for, 155
  sustainability and, 150–151
  urbanism and, 151
Biosphere
  cities and, 152–155
  history of concept, 44–46
  improvisation and, 201–203
  network archetypes, 93–98, 239
  visualization and, 101–103
Borders
  boundaries, 144
  controversy about, 159
  defined, 143
  ecological, 143
  invasive species and, 144
  open, 144–145
  territory and, 145–147
Browne, Janet, 89
Burghardt, Gordon M., 185

Carhart-Harris, Robin L., 210–211
Carson, Rachel, 102
Cheshire, James, and Oliver Uberti,
  123
Citizenship
  grassroots, 152
  place and, 148
Climate activists, 3–4

Cohen, Robin, and Olivia Sherrington, 156–157, 161
Collective spell, 63–66
Colonization, 141
Community-based democracy, 152
Connectivity
  concept of, 89–90, 106, 112
  forests and, 217–220
  improvisation and, 191–193
  local and global, 165
  long-term environmental change and, 192–193
Constructive connectivity, 111–112, 239
Consumerism
  collecting and, 80–81
  compulsion of, 63–66, 68–69
Contact zones, 155–158, 165, 239
Cosmopolitan bioregionalism, 139, 150, 159–162, 239
Cosmopolitan ideal, 152, 159
Crawford, Michael, 65–66, 75
Creolization, 156–157
Csikszentmihalyi, Mihaly, 189
Curate, 81–83, 110, 240
Cycles, 181

Darwin, Charles, 89, 102, 201
Deliberate pause, 76–78, 240
Deming, Alison, 217
Democracy
  improvisation and, 193–195
  inclusion and, 50–52
Developmental interludes
  computer screens, 57–60
  connect of, 14, 240
  examples of, 20
  family migration, 113–115
  lifecycle changes, 24–25
Difference, 155–158, 165
Displacement, 41–43, 52, 117–120
Diversity, 46–48, 164
Dodge, Jim, 13, 139
Dowie, Mark, 140–141, 146, 162

Ecocosmopolitan, 166
Ecological identity, 51, 240

Ecological knowledge gap, 11–12
Ecological networks, 98–101
Ecological urbanism, 151, 152–155
Ecosystem services, 9, 36, 44, 50, 240
Educational strategies. *See also* Adaptive learning; Developmental interludes; Perennial learning
  animal encounters and, 214–217
  cognitive dissonance, 23
  collaborative exploration of, 13
  edge of learning, 14
  educational narratives and, 29
  glimpses, 205–208
  I Ching, 199–201
  interpretation, 14
  language, 174–178
  meaningful conversation, 53
  methods, 31
  microbiomes, 220–223
  middle grounds, 211
  perspective taking, 56
  role of metaphor, 14
  role of the teacher, 5
  synthesis, 14
  thought experiments, 13
  visualization and, 98–103
Empathy, 122, 131–132
Entropic mind, 208–212
Environmental change makers, 10–11, 35–37
Environmental justice, 11, 50, 51
Environmental learning
  challenges of, 207
  concept of, 240
  future of, 28
  natural history observation, 76–78
  rationale for, 4, 9–10, 11
  strategies for, 13
  use of narratives, 56
Environmental studies programs, 7
Equity and inequality, 48–50
Evolutionary inheritance systems, 27

Ferguson, Niall, 87, 103
Filters, 78–79, 240
First Nations, 36, 140–142, 162
Flannery, Kent, and Joyce Marcus, 49

Floating bubbles, 246
Flow, 89–190, 191
Fluid sovereignties, 146–147, 240
Forbes, Peter, 151
Forman, Richard T.T., 100, 143

Generosity, 227–229
Ghosh, Amitav, 39, 57
Global diaspora, 125–126
Global environmental change, 39–41, 43–46, 62, 169, 180–182, 184, 192–193
Globalization, 124–126, 140, 146
Global Report on Internal Displacement, 41
Golden spike, 87–90, 241
Golley, Tristan, 181
Gratitude, 207
Greene, Joshua, 48, 147

Habitat fragmentation, 45
Hall, Stephen, 26, 137
Hall, Stuart, 155
Hanh, Thich Nhat, 117, 213
Harmon, Katherine, 102, 135
Haskell, David George, 94, 218–220
Hay, Pete, 142
Heise, Ursula, 161
Homophily, 104–105, 241
Horowitz, Alexandra, 225–226
Human behavior, 70–71
Human evolutionary diversity, 126, 129–131
Human Genome Project, 122
Humboldt, Alexander Von, 40, 102, 209

I Ching
  environmental change examples, 199–201, 246
  experiences with, 29–31
  improvisational knowledge system and, 195–199
Impey, Chris, 185
Improvisation
  biosphere and, 201–203
  creativity and, 190–191

definition, 170
environmental learning and, 169–171
etymology, 173
evolutionary origins of, 187–189
flow and, 189–191
Harold and the Purple Crayon, 21–23
language and, 174–178
networks and, 191–193
observation and, 178–180
random process and, 183–186
walking and, 172–174
Improvisational excellence, 241
Inclusion, 50–52
Indigenous, 163
Information visualization, 102
Intention, 71–73
Interbeing, 117, 213
Internet
  autonomic unfolding of, 63
  consumerism and, 63–66
  environmental perception and, 57–60

Jablonksi, Eva, and Marion J. Lamb, 28–29
Johnson, Steven, 106–107, 201
Jordan, Winthrop D., 129

Kadushin, Charles, 103–106
Kahneman, Daniel, 70
Klein, Naomi, 40
Knox, Paul, 127, 128
Kosala, J., 143

Laland, Kevin, 28–29
Language making nature, 176–178, 245
Lima, Manuel, 101–102
Liquid networks, 106, 241
Local and global, 165
Lukas, David, 176–178

Macfarlane, Robert, 174–176, 181
Maier, Charles S., 141, 145
Manahatta, 153

Mapping
  bioregionalism, 162–166
  difference, 165
  migration, 132–136, 244
  networks, 107–112
  sense of place, 137–140
  stories, 137–140, 244
Margulis, Lynn, 221, 222–223
Meeker, Joseph, 75
Memory
  bookshelves and, 59
  childhood, 19–24
  place and, 223–225
  reciprocity and, 225
Migrants, 117–120
Migration
  biosphere and, 122–124
  education and, 158
  environmental learning and, 120–122
  evolutionary globalization of, 123–124
  human globalization and, 124–126
  identity and, 156–157
  mapping of, 132–136
  nationalism and, 120
  public awareness and, 131–132, 156–158
  urbanization and, 127–129
Mindfulness, 76–78, 241
Minford, John, 29, 198–199
Ming-Dao, Deng, 197
Monadnock, 47, 172, 177, 183, 205
Montgomery, Sy, 214–215
Mycelia, 94, 97, 209, 219

Nachmanovitch, Stephen, 63, 185, 191, 193, 202
National Ecological Observatory Network, 192–193
Nationalism, 52
Natural history observation
  biosphere archetypes, 93–98
  collections and, 80
  ecological networks, 98–101
  migration, 133
  passing moments of, 76–78
  patterns, 180–182, 184
  reading the day and, 75–76
  Thoreau and, 73–75
Nelson, Richard, 216
Networks
  biosphere and, 93–98
  defined, 91, 241
  described, 89–90
  ecological, 92, 98–101
  education and, 107–112
  forests and, 217–220
  navigating, 107–111, 245
  observing, 90–93, 244
  personal identity and, 105–107
  social 92, 103–105
  visualizing of, 99, 101–103, 107–112
Network thinking, 91
Nielsen, Michael, 192

Obrist, Hans Ulrich, 81, 83

Pacific Northwest, 37–39, 149
Pagel, Mark, 46, 48, 70, 120, 125
Pattern-based environmental learning, 180–182, 183–184, 241, 245
Perceptual reciprocity
  animal encounters and, 214–217
  definition of, 207–208, 241
  forests and, 217–220
  memory and landscape, 223–225
  neighborhood and, 225–227, 246
Perennial learning, 26–27
Piketty, Thomas, 49
Place
  citizenship and, 148
  language and, 174–178
  migration and, 148–149
  sense of, 138, 148, 245
  urban, 149
Pollan, Michael, 208–211
Portal, 57–60, 241, 243
Powers, Richard, 19, 66, 91, 218–220, 223, 225, 226
Pratt, Mary Lois, 157
Precarity, 160

Pretor-Pinney, Gavin, 181
Proliferation, 78–79, 241

Random process, 106, 183–186
Reading the day, 75–76, 244
Reciprocity, 170
Redfern, Ron, 124
Refugees, 117–120
Reich, David, 125, 129, 130
Rothenberg, David, 169, 183, 188, 193, 216
Rovelli, Carlo, 12, 24, 91
Rutherford, Adam, 117, 130

Safina, Carl, 217
Sagan, Dorion, 213, 221, 222–223
Sanderson, Eric, 152–153, 223–224
Sapolsky, Robert, 47, 48–49, 55, 69, 71, 147
Saunders, Doug, 127
Savoy, Lauret E., 115
Scale, 109, 133
Schwenk, Theodore, 190
Snyder, Gary, 19, 51–52, 150, 171–172, 174
Sobel, David, 200–201
Social network analysis, 104–105
Solnit, Rebecca, and Joshua Jelly-Schapiro, 135, 137, 163
Species extinction, 44–45
Stamets, Paul, 94
Sustainable partnerships, 151

Thomashow, Mitchell
  Bringing the Biosphere Home, 40, 139
  Ecological Identity, 51
Thoreau, Henry David, 73–75, 81, 100, 102, 172, 174, 209
Thorpe, Helen, 158
Tides of change
  1960s and, 8
  educational challenge, 55–56
  four questions, 8–9
  historical challenge, 53–55
  Pacific Northwest, 37–39
Transboundary issues, 145

Transcontinental railroad, 87–90
Trentmann, Frank, 80
Tribalism
  concept, 46–48
  evolutionary origins of, 70–71
  global environmental change and, 48
  migration and, 121, 147
  moral behavior and, 48
  neurobehavioral research and, 47
  sovereignty and, 145–147
  territory and, 145–147
Tsing, Anna Lowenhaupt, 94, 160–161

Ubiquitous novelty, 66–67, 241
Umwelt, 212–214
UN Refugee Agency, 119, 131
Urban ecology, 129, 152–155
Urbanization, 127–129
Urban and rural, 38, 129

Valentine, Gill, 156
Vermeij, Geerat J., 23, 124, 186–187
Vigilance, 52–54
Visualization
  biosphere, 101–103
  ecological networks, 98–101
Volk, Tyler, 181
Von Uexkull, 212–214

Walls, Laura Dassow, 73
Wessels, Tom, 179–180, 182
White, Richard, 87
Whole Earth Catalog, 6–7, 138
Wilcove, David, 122–123, 124
Williams, Rosalind, 61, 80
Wilson, E. O., 44
World cities, 128–129
Workmen's Circle, 113–114

Yong, Ed, 220–222